Geraldine Rauch

The LORELIA Residual Test

Geraldine Rauch

The LORELIA Residual Test

A New Outlier Identification Test for Method Comparison Studies

Südwestdeutscher Verlag für Hochschulschriften

Impressum/Imprint (nur für Deutschland/ only for Germany)
Bibliografische Information der Deutschen Nationalbibliothek: Die Deutsche Nationalbibliothek verzeichnet diese Publikation in der Deutschen Nationalbibliografie; detaillierte bibliografische Daten sind im Internet über http://dnb.d-nb.de abrufbar.
Alle in diesem Buch genannten Marken und Produktnamen unterliegen warenzeichen-, markenoder patentrechtlichem Schutz bzw. sind Warenzeichen oder eingetragene Warenzeichen der jeweiligen Inhaber. Die Wiedergabe von Marken, Produktnamen, Gebrauchsnamen, Handelsnamen, Warenbezeichnungen u.s.w. in diesem Werk berechtigt auch ohne besondere Kennzeichnung nicht zu der Annahme, dass solche Namen im Sinne der Warenzeichen- und Markenschutzgesetzgebung als frei zu betrachten wären und daher von jedermann benutzt werden dürften.

Verlag: Südwestdeutscher Verlag für Hochschulschriften Aktiengesellschaft & Co. KG
Dudweiler Landstr. 99, 66123 Saarbrücken, Deutschland
Telefon +49 681 37 20 271-1, Telefax +49 681 37 20 271-0
Email: info@svh-verlag.de
Zugl.: Bremen, University, Dissertation, 2009

Herstellung in Deutschland:
Schaltungsdienst Lange o.H.G., Berlin
Books on Demand GmbH, Norderstedt
Reha GmbH, Saarbrücken
Amazon Distribution GmbH, Leipzig
ISBN: 978-3-8381-1310-4

Imprint (only for USA, GB)
Bibliographic information published by the Deutsche Nationalbibliothek: The Deutsche Nationalbibliothek lists this publication in the Deutsche Nationalbibliografie; detailed bibliographic data are available in the Internet at http://dnb.d-nb.de.
Any brand names and product names mentioned in this book are subject to trademark, brand or patent protection and are trademarks or registered trademarks of their respective holders. The use of brand names, product names, common names, trade names, product descriptions etc. even without a particular marking in this works is in no way to be construed to mean that such names may be regarded as unrestricted in respect of trademark and brand protection legislation and could thus be used by anyone.

Publisher: Südwestdeutscher Verlag für Hochschulschriften Aktiengesellschaft & Co. KG
Dudweiler Landstr. 99, 66123 Saarbrücken, Germany
Phone +49 681 37 20 271-1, Fax +49 681 37 20 271-0
Email: info@svh-verlag.de

Printed in the U.S.A.
Printed in the U.K. by (see last page)
ISBN: 978-3-8381-1310-4

Copyright © 2010 by the author and Südwestdeutscher Verlag für Hochschulschriften Aktiengesellschaft & Co. KG and licensors
All rights reserved. Saarbrücken 2010

Acknowledgments

This thesis would not have been possible without the helpful support, the motivating advises and the constructive guidance of my supervisors Prof. Dr. Jürgen Timm, Dr. Christoph Berding and Dr. Andrea Geistanger to whom I owe my special gratitude.

Moreover, I want to thank the whole department of Biostatistic of Roche Diagnostics Penzberg, as I have never worked in a more friendly and helpful atmosphere.

My special thanks go to my wonderful family, who always encouraged and supported me.

Finally, I want to thank my friend Peter Gebauer, who helped me to keep my calm.

Geraldine Rauch

Contents

1 **Introduction**   7

2 **Overview of the Theory of Outliers**   10
    2.1 History of Research 10
    2.2 Motivation of Outlier Identification and Robust Statistical Methods 11
    2.3 An Informal Definition of Outliers 12
       2.3.1 Outliers, Extreme Values and Contaminants 13
       2.3.2 The Diversity of Extremeness 18
           2.3.2.1 Extremeness with Respect to the Majority of Data 18
           2.3.2.2 The Importance of Underlying Statistical Assumptions 19
           2.3.2.3 Extremeness in Multivariate Datasets 20
           2.3.2.4 Ambiguity of Extreme Values 22
    2.4 A Short Classification of Outlier Candidates 23
       2.4.1 The Statistical Assumptions 24
       2.4.2 Causes for Extreme Values 24
       2.4.3 Different Goals of Outlier Identification 24

3 **Different Concepts for Outlier Tests**   26
    3.1 Classification of Outlier Tests 26
       3.1.1 Tests for a Fixed Number of Outlier Candidates 26
       3.1.2 Tests to Check the Whole Dataset 27
    3.2 Formulation of the Test Hypotheses 29
       3.2.1 Discrepancy Tests 29
       3.2.2 Incorporation of Outliers 30
           3.2.2.1 The Inherent Hypotheses 30
           3.2.2.2 The Deterministic Hypotheses 30

		3.2.2.3	The Mixed Model Alternative	31

	3.3	Problems and Test Limitations	31
		3.3.1 The Masking Effect	32
		3.3.2 The Swamping Effect	32
		3.3.3 The Leverage Effect	33

4 Evaluation of Method Comparison Studies — 36

4.1	Comparison by the Method Differences	36
	4.1.1 The Absolute Differences	37
	4.1.2 The Relative Differences	38
4.2	Comparison with Regression Analysis	42
	4.2.1 Robust Regression Methods	43
	4.2.1.1 Deming Regression	44
	4.2.1.2 Principal Component Analysis	45
	4.2.1.3 Standardized Principal Component Analysis	45
	4.2.1.4 Passing-Bablok Regression	47

5 Common Outlier Tests for MCS — 49

5.1	Outlier Tests based on Method Differences	50
	5.1.1 Problems and Limitations	51
5.2	Outlier Test based on Regression	51
	5.2.1 Problems and Limitations	53

6 The New LORELIA Residual Test — 54

6.1	Statistical Assumptions for the New Test	55
6.2	The Concept of Local Confidence Intervals	56
6.3	How to Weight - Newly Developed Criteria	58
	6.3.1 Historical Background - Basic Ideas	58
	6.3.1.1 Problems and Limitations	59
	6.3.2 New Concepts for Weight Construction	63
	6.3.2.1 Construction of a Local Estimator	63
	6.3.2.2 Construction of an Outlier Robust Estimator	63
	6.3.2.3 Invariance under Axes Scaling	64
	6.3.2.4 The Meaning of the Local Data Information Density	64
	6.3.2.5 The Co-Domain of the Weights	65

6.4	The Weights for the LORELIA Residual Test		65
	6.4.1 Definition of the Distance Measure		66
	6.4.2 Definition of a Reliability Measure		70
6.5	Definition of the LORELIA Residual Test		72

7 Performance of the New LORELIA Residual Test — 75

- 7.1 The LORELIA Residual Test in Comparison to Common Outlier Tests 77
  - 7.1.1 Performance Comparison for Real Data Situations 78
    - 7.1.1.1 No Suspicious Values 78
    - 7.1.1.2 One Outlier Candidate 82
    - 7.1.1.3 Uncertain Outlier Situation 85
    - 7.1.1.4 Decreasing Residual Variances 88
    - 7.1.1.5 Very Inhomogeneous Data Distribution 90
    - 7.1.1.6 Conclusion 94
  - 7.1.2 Proof of Performance Superiority for an Exemplary Data Model 94
  - 7.1.3 Performance Comparison for Simulated Datasets 101
    - 7.1.3.1 Simulation Models 102
    - 7.1.3.2 Evaluation of the Simulation Results 105
      - Actual Type 1 Error Rates 106
      - True Positive and False Positive Test Results 107
    - 7.1.3.3 General Observations and Conclusions 113
- 7.2 Influence of the Outlier Position on its Identification 115
  - 7.2.1 Simulation Models 116
  - 7.2.2 Homogeneous Data Distribution 118
    - 7.2.2.1 Constant Residual Variance 118
      - Expected Results 118
      - Observed Results 121
    - 7.2.2.2 Constant Coefficient of Variance 124
      - Expected Results 124
      - Observed Results 125
  - 7.2.3 Inhomogeneous Data Distribution 126
    - 7.2.3.1 Constant Residual Variance 127
      - Expected Results 127
      - Observed Results 127

		7.2.3.2 Constant Coefficient of Variance	130

                          Expected Results 130

                          Observed Results 130

    7.3   How to Deal with Complex Residual Variance Models 133

    7.4   Considerations on the Alpha Adjustment 136

    7.5   Summary of the Performance Results 137

8   Conclusions and Outlook   140

A   Software Development and Documentation   145

B   Test Results of Section 7.1.3   148

    B.1   Constant Residual Variance 148

    B.2   Constant Coefficient of Variance 162

    B.3   Non Constant Coefficient of Variance 176

Symbols   190

Bibliography   192

Chapter 1

Introduction

In this work, a new outlier identification test for method comparison studies based on robust linear regression is proposed in order to overcome the special problem of heteroscedastic residual variances.

Method comparison studies are performed in order to prove equivalence or to detect systematic differences between two measurement methods, instruments or diagnostic tests. They are often evaluated by linear regression methods. As the existence of outliers within the dataset can bias non robust regression estimators, robust linear regression methods should be preferred. In this work, the use of Passing-Bablok regression is suggested which is described in [Passing, Bablok, 1983], [Passing, Bablok, 1984] and [Bablok et al. 1988]. Passing-Bablok regression is a very outlier resistant procedure which takes random errors in both variables into account. Moreover, the measurement error variances are not required to be constant, so Passing-Bablok regression is still appropriate if the error variances depend on the true concentration which is a common situation for many laboratory datasets.

Beside the use of robust regression methods, it is strongly recommended to scan the dataset for outliers with an appropriate outlier test, as outliers can indicate serious errors in the measurement process or problems with the data handling. Therefore, outliers should always be carefully examined and reported in order to detect possible error sources and to avoid misinterpretations.

If method comparison is evaluated by a robust regression method (here Passing-Bablok), outliers will correspond to measurement values with surprisingly large orthogonal residuals. A possible approach for the identification of outliers is the construction of confidence intervals for the orthogonal residuals which will serve as outlier limits. These confidence intervals will depend on the underlying residual variance, which has to be estimated. Note that only robust variance estimators are appropriate in this context as otherwise existing outliers will bias the estimate.

Common outlier tests for method comparison studies are based on global, robust outlier limits for the residuals of a regression analysis or for the measurement distances. In the work of [Wadsworth, 1990], global, robust outlier limits of the form

$$\mathrm{med}(\cdot) \pm q \cdot \mathrm{mad68}(\cdot)$$

are proposed, where q correspond to some predefined quantile. This approach can be applied to any of the comparison measures proposed above. However it requires that the measurement error variances or the residual variances, respectively, remain constant over the measuring range. If the variances follow a simple model, for example if they are proportional to the true concentration (constant coefficient of variance) the same concepts can be applied after an appropriate data transformation. However, in many practical applications the error variances or residual variances, respectively, do not follow a simple model - they are neither constant nor proportional to the true concentration and the underlying variance model is unknown. In this case none of the transformation methods proposed in the literature will fit and common robust variance estimators as proposed in [Wadsworth, 1990] will not be appropriate.

The new LORELIA Residual Test (=LOcal RELIAbility) is based on a local, robust residual variance estimator $\hat{\sigma}_{r_i}^2$, given as a weighted sum of the observed residuals r_k. Outlier limits are given as local confidence intervals for the orthogonal residuals. These outlier limits are estimated from the actual data situation without making assumptions on the underlying residual variance model. The local residual variance estimator for the ith orthogonal residual is given as the sum of weighted squared residuals r_k^2:

$$\hat{\sigma}_{r_i}^2 = \frac{1}{\sum_{l=1}^n w_{il}} \cdot \sum_{k=1}^n w_{ik} \cdot r_k^2, \quad \text{for } i = 1, ..., n.$$

The LORELIA Weights w_{ik} are given as:

$$w_{ik} := \Delta_{ik} \cdot \Gamma_{k,n}, \quad \text{for } i, k = 1, ..., n,$$

where Δ_{ik} is a measure for the distance between r_i and r_k along the regression line to ensures that the residual variance is locally estimated and $\Gamma_{k,n}$ is a measure for the local reliability to guarantee that the residual variance estimator is robust against outliers.

The present work is organized as follows:

In Chapter 2, a general overview of the theory of outliers is given. The relation between outlier identification and robust statistical methods is discussed. Moreover, an informal definition of the expression 'outlier' is given. Finally, a classification for different outlier scenarios is proposed.

In Chapter 3, different concepts for outlier tests are presented based on the work of [Hawkins, 1980] and [Barnett, Lewis, 1994]. A classification of outlier tests is given and different kind of test hypotheses are presented. Moreover common problems and limitations which can complicate the identification of outliers are discussed.

Different approaches for the evaluation of method comparison studies are presented in Chapter 4. The comparison of two measurement series can be either done by analyzing the differences between the measurement values (compare references [Altman, Bland, 1983], [Bland, Altman, 1986], [Bland, Altman, 1995] and [Bland, Altman, 1999]) or by fitting a linear regression line as described in [Hartmann et al. 1996], [Stökl et al., 1998], [Linnet, 1998] and [Linnet, 1990]. Both concepts are discussed.

Common outlier tests for method comparison studies and their limitations are presented in Chapter 5. These tests which are proposed by [Wadsworth, 1990] are based on global, robust outlier limits for the residuals of a regression analysis or for the measurement distances, respectively.

The new LORELIA Residual Test is introduced in Chapter 6. After the presentation of the general concepts for local confidence intervals, the requirements for an appropriate weighting function are discussed. Finally, the LORELIA Residual Test is explicitly defined.

In Chapter 7, the performance of the LORELIA Residual Test is evaluated based on different criteria. To begin with, it will be checked visually if the new test identifies surprisingly extreme values truly as outliers and if it performs better than the standard outlier tests presented in Chapter 5. Subsequently, the superiority of the LORELIA Residual Test is theoretically proven for datasets belonging to a simple model class M. Based on a simulation study, all test are compared with respect to the number of true positive and false positive test results. As the LORELIA Residual Test is a *local* outlier test, the identification of an outlier depends on its position within the measuring range. Another simulation study is performed in order to evaluate the influence of the outlier position within the dataset on its identification. As the outlier test corresponds to a multiple test situation, the local significance levels have to be adjusted. Different adjustment procedures and their properties are discussed. Finally, performance limitations of the new test are presented. The LORELIA Residual Test is only appropriate if the local residual variances do not change too drastically over the measuring range and if the sample distribution is not too inhomogeneous. This problem is discussed and a solution is suggested.

A summary of this work is given in Chapter 8. Open questions and suggestions how to handle them will be presented in an outlook.

Chapter 2

Overview of the Theory of Outliers

The theory of outliers is split in many different research areas in our days. Outliers have been mentioned in statistical contexts for centuries as the problem how to deal with extreme observations is a very intuitive one. This chapter will give an introduction to the statistical theory of outliers.

In Section 2.1, the early history of statistical research on outliers is briefly presented. In Section 2.2, the relationship of outlier identification and robust statistical methods in data analysis is discussed. In Section 2.3, an informal definition for the expression 'outlier' is determined. Finally, in Section 2.4, a classification for different outlier scenarios is given.

2.1 History of Research

The subject of outliers in experimental datasets has been broadly and diversely discussed in the statistical literature for centuries. In this section, a brief history of the early beginnings of outlier theory will be given.
Informal descriptions of outliers and how to handle them go back to the 18th century. A first discussion of the problem if outliers should be excluded from data analysis was given by [Bernoulli, 1777] in the context of astronomical observations.
[Peirce, 1852] was the first to publish a rather complicated test for outlier identification based on the assumption of a mixed distribution describing the normal and the outlying data.
A more intuitive test for the identification of a single outlier was presented by [Chauvenet, 1863]. Assuming that the sample population follows a normal distribution $N(0, \sigma^2)$, his test is based on the fact that the expected number of observations exceeding $c \cdot \sigma$ in a sample of size n is given by $n \cdot \Phi(-c)$, where $\Phi(\cdot)$ is the distributional function of the standard normal distribution. He proposed to reject any observation which exceeds $c \cdot \sigma$ where c fulfills $n \cdot \Phi(-c) = 0.5$. Hence, the test is expected to reject half an observation of the normal data per sample, regardless of the sample size n. Thus the probability to reject any observation as an outlier

in a sample of size n is given by $\frac{1}{2n}$. The chance of wrongly identifying at least one normal data value as an outlier is hence given by $1 - \left(1 - \frac{1}{2n}\right)^n$ which increases with the sample size and becomes unreasonably large. The concepts of [Chauvenet, 1863] were further developed and varied by [Stone, 1868].

Several rejection tests for outliers based on the studentized measurement values were proposed in the following years by different authors (compare e.g. [Wright, 1884]). The studentized values are a transformation of the original values given by $\frac{x_i - \bar{x}}{s_{xx}}$, where $\bar{x}$ is the mean and s_{xx} is the empirical standard deviation of the measurement values $x_1, x_2, ..., x_n$.

[Goodwin, 1913] proposed to exclude the identified outliers in the calculation of the sample mean and the sample standard deviation. Years later [Thompson, 1935] showed however, that this modified test is a monotonic function of the original test. [Thompson, 1935] was also the first who constructed an exact test for the test statistic $\frac{X_i - \bar{X}}{S_{xx}}$.

[Irwin, 1925] was the first to propose an outlier test based on a test statistic involving only extreme values. For the ordered sequence $X_{(1)} \leq X_{(2)} \leq ... \leq X_{(n)}$, he used the test statistic $\frac{X_{(n-k+1)} - X_{(n-k)}}{S_{xx}}$ in order to test if the k most extreme values are outliers.

Finally [Pearson, Sekar, 1936] found out some important results on the underlying significance level for the test based on the studentized test statistic $\frac{X_i - \bar{X}}{S_{xx}}$. They also were the first to discuss the 'masking effect' which will be presented in Section 3.3.1.

Since these times, many important publications in the field of outlier theory were made. However, the diversity and complexity of outlier scenarios has increased immensely, so a general overview of research results for all areas of outlier theory will not be possible in this context. In the following sections, additional authors will be cited with respect to the subjects related to the topic of this work.

2.2 Motivation of Outlier Identification and Robust Statistical Methods

Experimental datasets sometimes contain suspicious extreme observations which do not match to the main body of the data. These values can bias parameter estimates and thus influence the evaluation of data analysis. In order to avoid this, there exist two approaches:

1. The use of robust statistical methods,
2. The identification of so called 'outliers' before any data analysis.

Most robust non parametric methods replace the numerical values by their respective ranks. However, the numerical values of outliers and extreme observations are important to judge the stability of the measurement process and they can give valuable information about the underlying model or distribution of the dataset. Outliers can indicate possible error sources and they may motivate the data analyst to adjust his statistical assumptions. Therefore, the identification of outliers is an important part of data analysis which can not entirely be replaced by robust methods as robust methods involve a certain loss of data information.

A special robust approach to protect against outliers is the use of 'α-trimming'. Here, the upper and lower $\alpha\%$ of values are deleted before any data analysis. This will stabilize estimators in models or distributions since existing outliers will be deleted. The size of α will determine the 'degree of robustness' but also the 'degree of information loss'. For the extreme case of $\alpha = 50\%$ the dataset is shrinkend to its median. A more detailed description of this method can be found in [Barnett, Lewis, 1994].

Note that, in many practical applications robust methods and outlier identification can not be regarded as alternatives. Data analysis is often based on both approaches. For example, outlier identification tests are often based on model and distributional assumptions with robustly estimated parameters. An short overview of robust statistical methods and its relation to outlier identification is given in [Burke, 1999].

2.3 An Informal Definition of Outliers

There exist no consistent mathematical definition of the term 'outlier' within the literature. Moreover, the expression is often used without a proper specification of its meaning. Therefore it is important to fix an informal definition before dealing with the specific outlier situations considered in this work.

Outlying observations can occur in any kind of data sample. The judgment on what kind of measurement value can be interpreted as an 'outlier' is often done in a very intuitive way depending on the structure of the data, the graphical presentation and the subjective impression of the data analyst. The following graphs visualize three completely different data situations and presentations. The outlying observations which are marked with an arrow or a circle, respectively, all have in common that they are 'surprisingly extreme values' with respect to the rest of the dataset. Extremeness is always related to the question what the analyst expects to observe. In Figure 2.1 all bars except the second are of similar height. In this case, all bars are expected to have a height about 20 units.

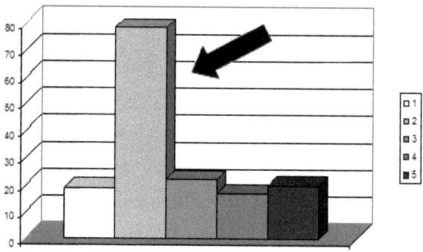

Figure 2.1: Outliers in Different Data Situations - Bar Diagram

In Figure 2.2, all data except one value seem to follow a linear model:

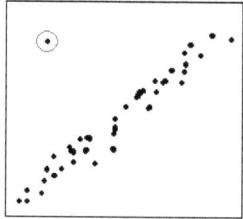

Figure 2.2: Outliers in Different Data Situations - Linear Model

The data points in Figure 2.3 may represent a data sample from a normal distributed population. Observations accumulate in the middle of the measuring range. Only one isolated value at the boundary seems suspicious.

Figure 2.3: Outliers in Different Data Situations - Normal Distribution

2.3.1 Outliers, Extreme Values and Contaminants

Outlying observations do not fit with the statistical assumptions which describe the majority of data. They belong to a different population and thus follow different statistical models or distributions. In order to describe a dataset which contains observations of several populations, the following notations will be used for convenience:

Definition 2.1 (Contaminants, Contaminated and Contaminating Population)
Consider a data sample of size N which should be representative for a given population P_{int} of interest. Suppose that $N_{\text{cont}} < N$ of the data values correspond to a different population $P_{\text{cont}} \neq P_{\text{int}}$. Then these N_{cont} values are called <u>contaminants</u> and the corresponding population P_{cont} is called <u>contaminating population</u> with respect to the <u>contaminated population</u> P_{int}. The given data sample thus represents a mixture of the populations P_{int} and P_{cont} rather than P_{int}.

The mixture of populations will now be defined mathematically:

Definition 2.2 (Mixed Distribution/ Model)
Consider a data sample S of size N which should represent the population P_{int} and which is contaminated by the contaminating population P_{cont}. Suppose that $P_{\text{int}} \sim F$ and $P_{\text{cont}} \sim G$ for two statistical distributions (or statistical models) F and G with $F \neq G$. Let p be the probability to choose an observation which belongs to P_{int}. Then, the data sample S is a realization of the <u>mixed distribution</u> (or the <u>mixed model</u>)

$$p \cdot F + (1-p) \cdot G.$$

Note that in practical applications p is usually close to 1. The aim is to identify the data which belongs to the contaminating population P_{cont}. The problem lays in separating the contaminants from the observations of interest. Often this is not entirely possible however.
In general it may be possible that the population of interest P_{int} is contaminated by several contaminating populations P_{cont_i} for $i = 1, ..., m$. The separation of several subpopulations is related to the field of cluster analysis and will not be further discussed here. For the sake of simplicity, in this work the problem will be reduced to the case of one contaminating population P_{cont}.

To illustrate the problem, consider the case where F and G are two different distributions. In the following graphical examples, it is assumed that:

$$F \sim N(\mu_1, \sigma_1^2)$$
$$\text{and} \quad G \sim logN(\mu_2, \sigma_2^2).$$

The distributions F and G can be separated best if they differ by a substantial shift of mean. In the following example, the probability p is given by

$$p := 0.9$$

and the distribution parameters are chosen as follows:

$$\begin{aligned} \mu_1 &:= 3, \\ \mu_2 &:= 7, \\ \sigma_1^2 &:= 1, \\ \sigma_2^2 &:= 1. \end{aligned}$$

Since the contaminating population P_{cont} has a much higher mean than the population of interest P_{int}, observations corresponding to large values are more likely to belong to the contaminants than to the population of interest. On the other hand, observations corresponding to small values are more likely to belong to the population of interest. Note that, contaminants with a small magnitude will be hidden within the samples which belong to P_{int}. However, the probability to observe a contaminant with a small value is very low, so this problem may be neglected.

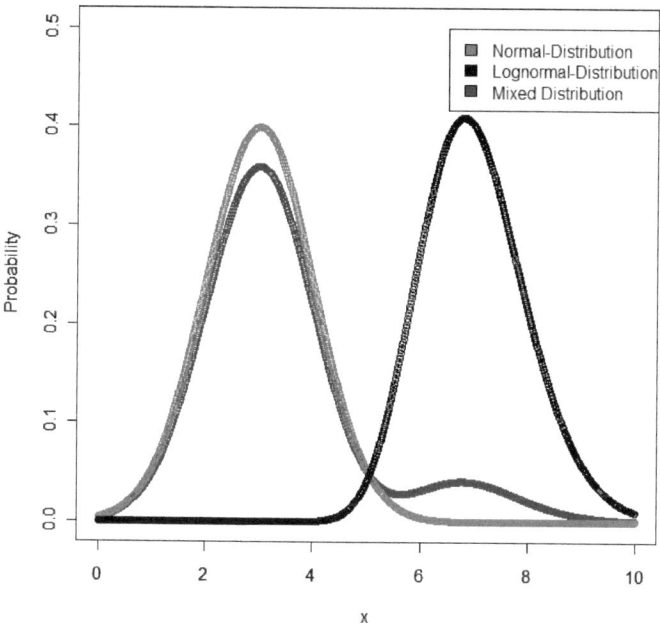

Figure 2.4: Mixed Distribution: $0.9 \cdot N(3,1) + 0.1 \cdot logN(7,1)$

Separation becomes more difficult if there is no shift in mean but in variance. For

$$p := 0.9,$$

choose:

$$\mu_1 := 5,$$
$$\mu_2 := 5,$$
$$\sigma_1^2 := 1,$$
$$\sigma_2^2 := 2.$$

Here, most of the observations of the contaminating *and* the contaminated population will have values close to the common mean $\mu_1 = \mu_2 = 5$. Those observations can not adequately be assigned to one of the populations. Only extreme values with very small or very high magnitudes are more likely to belong to the contaminants than to the population of interest.

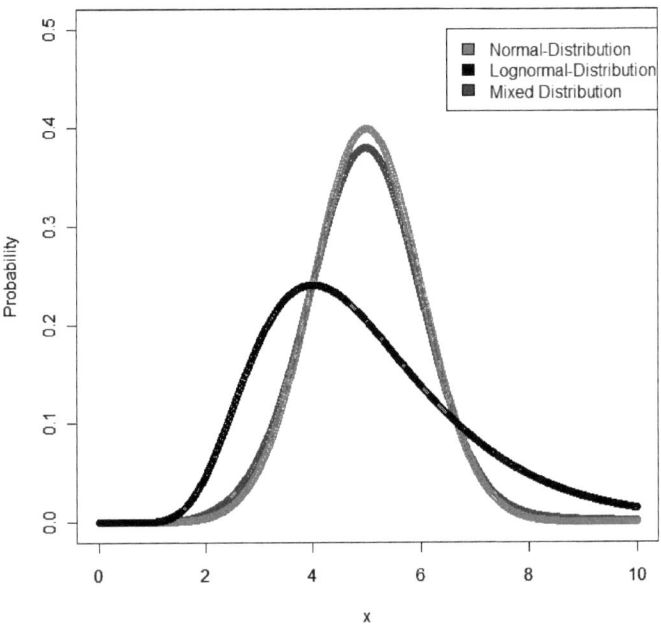

Figure 2.5: Mixed Distribution: $0.9 \cdot N(5,1) + 0.1 \cdot logN(5,2)$

It has been shown in examples that contaminants may be identified because of their extremeness. However, they may as well be completely hidden in the population of interest. This fact will lead to an informal definition of the expression 'outlier':

Notation 2.3 (Outlier)

An observation of a dataset S will be referred as an <u>outlier</u>, if it belongs to a contaminating population P_{cont} *and* if it is surprisingly extreme with respect to the model or distributional assumptions for the population of interest P_{int}.

In practical applications, the true population affiliation of a surprisingly extreme value is usually not known. Therefore, the following notation is introduced:

Notation 2.4 (Outlier candidate)

An observation of a dataset S will be referred as an <u>outlier candidate</u>, if it is surprisingly extreme with respect to the model or distributional assumptions for the population of interest P_{int}.

As the population affiliation usually can not be determined, this work will refer to the term 'outlier candidate' most of the time. Note that in the literature the above notations are not consistent.

The expressions 'outlier' and 'outlier candidate' can be defined mathematically by fixing a measure for 'surprisingly extreme'. Usually, this will be done by formulating hypotheses for an appropriate outlier test. The following remark will give an overview of the relations between the different notations given in this section:

Remark 2.5

(i.) By the Definitions 2.1, 2.2 and Notation 2.3, outliers are a subset of the contaminants.

(ii.) By Notation 2.3 and 2.4, outliers are a subset of the outlier candidates.

(iii.) Outlier candidates are not necessarily outliers. They may as well correspond to the population of interest!

(iv.) Contaminants are not necessarily outlier candidates. They can be hidden in the population of interest!

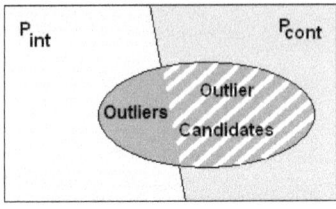

Figure 2.6: Population Affiliations

The aim of outlier tests is to separate the true outliers from the population of interest. This can only be successful if the population of interest P_{int} is well separated from the contaminating population P_{cont}. Most outlier identification rules only test the extremeness of observations with respect to the distribution of the population of interest P_{int} without making assumptions on the contaminating distribution or the mixed model parameter p. This approach was follow for example by [Davies, Gather, 1993] who defined so called "'outlier regions'" based on the distributional assumptions for the populations of interest.

Other outlier tests are based on special mixed model assumptions, which is strongly related to the theory of cluster analysis as mentioned above. Early research on outlier theory with respect to mixed model assumptions was done by [Dixon, 1950] and [Grubbs, 1950]. They were followed by [Anscombe, 1960], [Tukey 1960], [Box, Tiao, 1968], [Guttman, 1973], [Marks, Rao, 1979], [Aitkin, Wilson, 1980] and many others.

2.3.2 The Diversity of Extremeness

As it has been pointed out in the previous section, a measure for extremeness has to be defined in order to construct outlier tests and to define the term 'outlier' mathematically. However, the question of what should be considered as 'extreme' is not obvious. In this chapter, the most important considerations and remarks about extremeness of data values will be concluded.

2.3.2.1 Extremeness with Respect to the Majority of Data

In many data situations extreme observations will correspond to values of very high or very low magnitude. For most statistical distributions, data points accumulate around the mean value. Indeed, extreme values do not necessarily correspond to extremely small or big values. Extremeness is rather correlated to the *isolation* of the observation.

Consider for example the following graph of an U-distributed dataset. The majority of values is accumulated at the two boundaries. One extreme value is observed, marked with an arrow, which is close to the mean of the distribution:

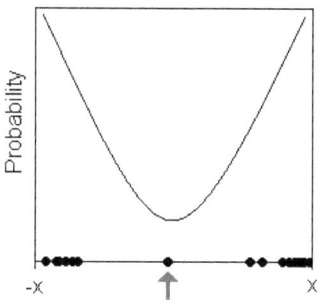

Figure 2.7: Extreme Observation for the U-Distribution

2.3.2.2 The Importance of Underlying Statistical Assumptions

A measure for the extremeness of observations will be determined by the statistical assumptions on the dataset. The other way round, wrong statistical assumptions can lead to wrong conclusions about extreme and non extreme values. For example, surprisingly extreme values for a normal distribution may not be considered as extreme under a more heavy tailed distribution like Student's-t.

Wrong assumptions on the data model can cause errors in the interpretation of extreme values as well. In the following graphical example, the dataset is wrongly assumed to be described by a linear regression model.

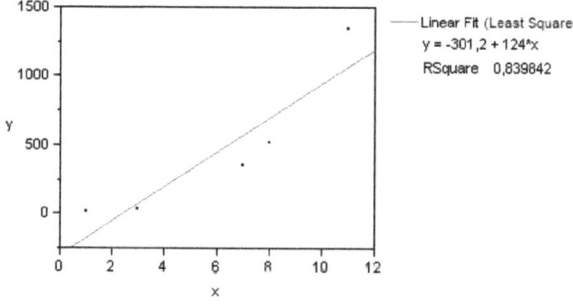

Figure 2.8: Error in the Model Assumption

Several residuals seems extremely high with respect to the regression line. A polynomial of degree 3 however fit the data almost perfectly and no extreme observation can be identified.

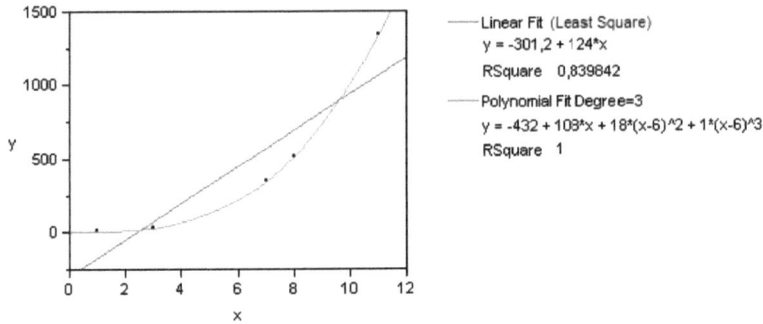

Figure 2.9: Corrected Model Assumption

2.3.2.3 Extremeness in Multivariate Datasets

Extreme values in multivariate datasets are much less obvious to identify than in univariate datasets. A visual inspection of the dataset is difficult since there often exist no easy way for a graphical representation. A discussion of this problem can be found in [Buttler, 1996].

A multivariate observations may contain extreme values in the single variables. However, an extreme value in one variable does not necessarily mean that the corresponding multivariate observation is extreme with respect to the underlying multivariate statistical distribution or model. The other way round, a multivariate observation may look surprisingly extreme with respect to the stated distribution or model whereas the values of the single variables all are just slightly shifted. Consider for example the following three dimensional dataset:

Obs	x_1	x_2	x_3
1	4	2	8
2	2	1	4
3	7	1	9
4	1	2	5
5	5	1	7
6	4	12	28
7	3	3	9
8	2	4	2

Table 2.1: Example for a Multivariate Dataset

Here, observation 6 is surprisingly extreme in the variables x_2 and x_3. However, all data values expect the 8th observation are perfectly fitted by the two-dimensional regression model $x_3 = x_1 + 2 \cdot x_2$. The 8th observation contains no extreme values within the single variables although it is an obvious outlier candidate:

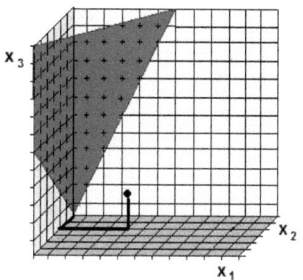

Figure 2.10: Outlier Candidate from a Two-Dimensional Linear Regression Model

Exemplary methods for the identification of multivariate outliers are discussed in [Acuña, Rodriguez, 2005].

If several groups of data are compared in a multivariate dataset, outliers can appear with respect to scale and location measures:

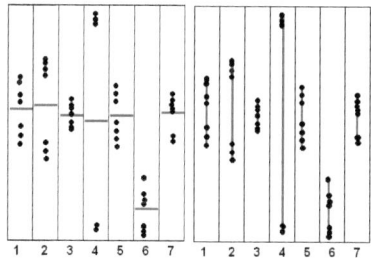

Figure 2.11: Outlier Candidates in Location and in Variance

Group 6 is an outlier candidate in location since the group mean differs significantly from all other group means. If however the variance of data within a single group is considered, group 4 turns out to be surprisingly extreme.

The principles of scale and location outliers are also discussed in [Burke, 1999]. Examples for corresponding scale and location outlier tests are given in [Wellmann, Gather, 2003].

2.3.2.4 Ambiguity of Extreme Values

Extreme observations in a dataset may be ambiguous. The following data are described by a linear regression model. There exist two suspicious observations, marked in gray and black, but it is not obvious which one of them is an outlier or if maybe both values are outliers.

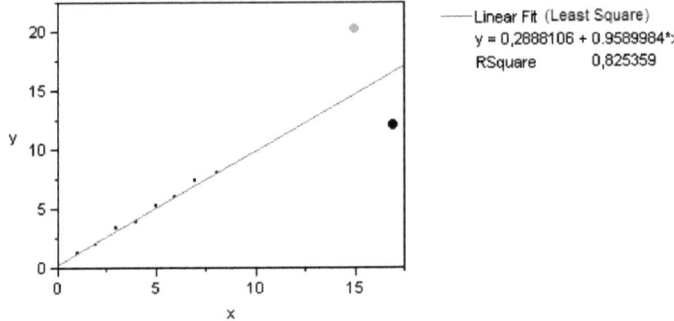

Figure 2.12: Ambiguity of Extreme Values

The model adjustment is not very satisfying with $R^2 = 0.825359$. If the first or the second suspicious value is removed, the corresponding linear fit becomes substantial different. As it can be deduced from the following graphs, both R^2 values are much higher now and approximately of the same magnitude. However, the parameter estimates are very different! Without further data information, it can not be deduced which observation is spurious or if even both values are outliers.

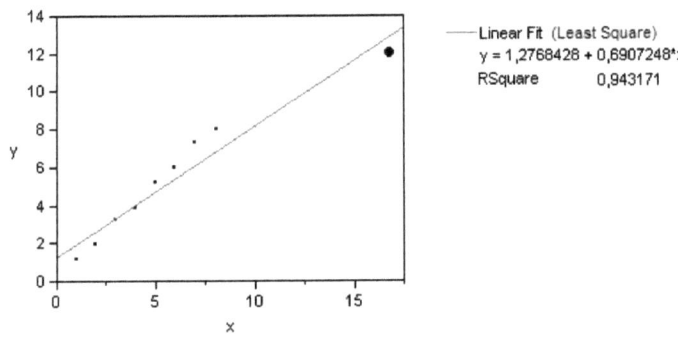

Figure 2.13: Linear Fits for Excluded Upper Extreme Value

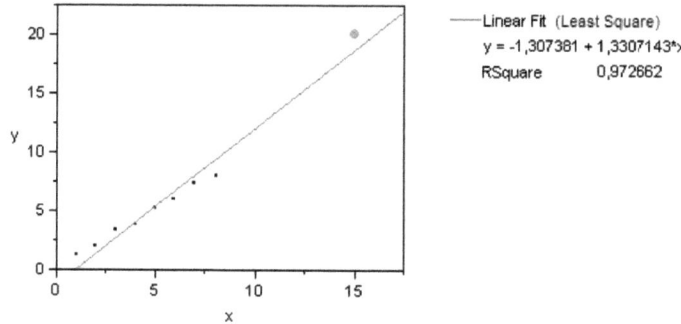

Figure 2.14: Linear Fits for Excluded Lower Extreme Value

If both suspicious values are removed, the fit is given by:

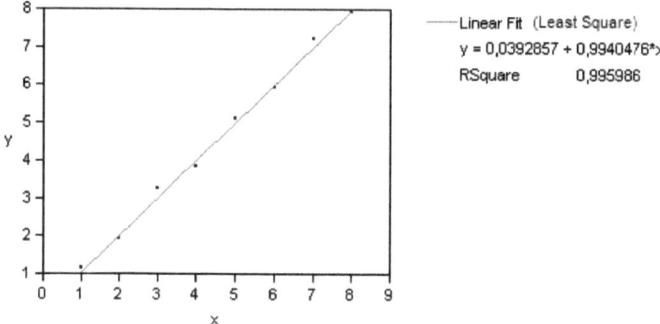

Figure 2.15: Linear Fits with both Extreme Values Excluded

With $R^2 = 0.995986$ the model adjustment has improved a lot. The parameter estimates are again very different from the previous ones. This points out that the R^2-value as a measure of fit can lead to serious misinterpretations.

2.4 A Short Classification of Outlier Candidates

There exist a variety of different outlier scenarios which differ concerning the structure of the dataset, the underlying statistical assumptions and the specific interests of the data analyst. It will be nearly impossible to define detailed subgroups for all possible outlier scenarios within an overall classification. The aim of this section is to give a classification which points out the most fundamental differences between the existing outlier scenarios.

As it has been mentioned in Remark 2.5 (iii.), surprisingly extreme values do not always belong to the contaminating population P_{cont}. Since in practical application the true population affiliation of the extreme value is not known, this section will present a classification for outlier candidates rather than for true outliers (compare Notation 2.4).

2.4.1 The Statistical Assumptions

In a first classification step the model or distributional assumptions for the population of interest P_{int} are considered. As it has been mentioned in Section 2.3.2.2, inappropriate statistical assumptions can lead to wrong conclusion on outlier candidates. Therefore, it should be verified if the outlier candidate is judged with respect to the right statistical assumptions.

2.4.2 Causes for Extreme Values

In the second classification step, outlier candidates will be divided into outliers, which belong to the contaminating population P_{cont} and extreme observations which are valid members of the population of interest P_{int}. In other words, the outlier candidates are classified concerning the cause for their extremeness. True outliers are due to the fact that the population of interest P_{int} really *is* contaminated. Extreme values which do not belong to the contaminants are due to the natural variance in the population of interest P_{int}. In this case, the outlier candidate provides a part valid information on the population of interest.

2.4.3 Different Goals of Outlier Identification

The last step in the classification of outliers will be determined by the predefined goal for the outlier identification, which will influence the formulation of hypothesis for the outlier test. If the outlier candidate is due to an error in the statistical assumptions for the population of interest, the aim will be to adjust these assumptions. After an appropriate adjustment of the statistical model, the outlier candidate becomes a regular member of the population of interest. If the outlier candidate really is a contaminant, the causes for the contamination should be explored and removed whenever possible. A new measurement under corrected conditions can replace the outlying value. Care have to be taken with the identification of contamination causes since identifying wrong causes may effect the results of data analysis. If the outlier candidates belongs to the population of interest it should not be removed since they involve valid information about the underlying distribution.

In most cases however, it is not easy or even impossible to decide whether an outlier candidate is due to the natural variation in the population of interest or if it is due to contamination or if it reflects a misconception in the statistical modeling.

A supplementary possibility to deal with outlier candidates which will not be further discussed here is 'accommodation' as referred in [Barnett, Lewis, 1994]. This can be done by 'Winsorization' where outlier candidates are replaced by their nearest neighbors.

The following flow chart will visualize the different steps of outlier classification.

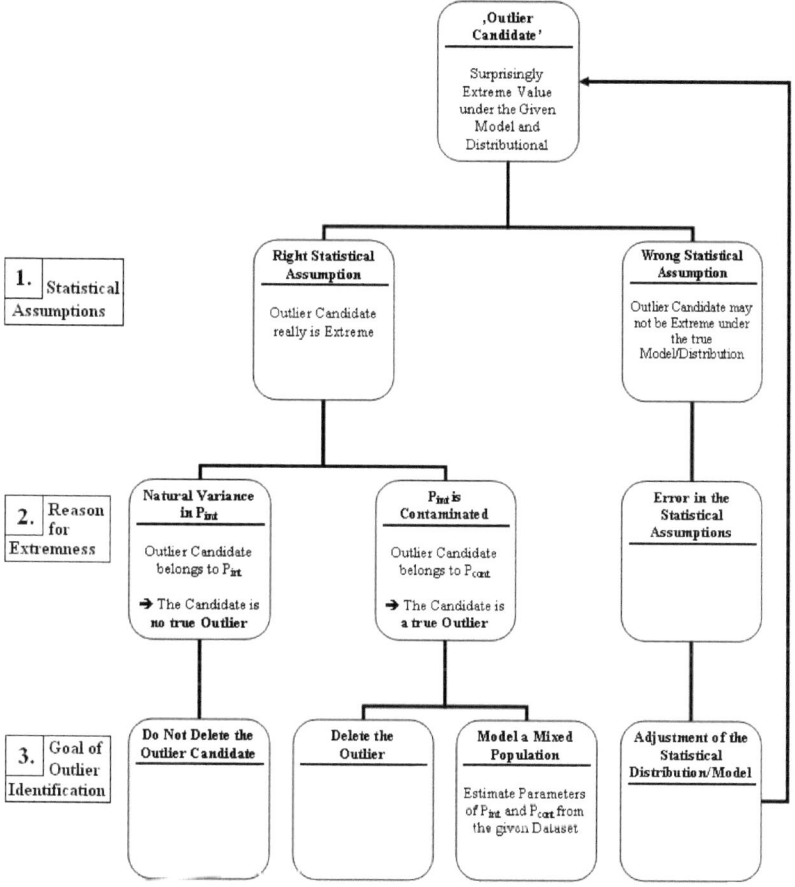

Figure 2.16: Classification of Outlier Candidates

Chapter 3

Different Concepts for Outlier Tests

The identification of outlier candidates as motivated in Chapter 2 will be based on statistical tests. Thereby, the diversity of outlier scenarios correspond to a broad field of different outlier tests. In this chapter, several types of outlier test will be presented. In Section 3.1, a short classification of different outlier tests will be given whereas Section 3.2 describes different types of test hypotheses. Finally, in Section 3.3, some basic problems which can be met in the identification of outliers are presented.

With the notations introduced in Section 2.3.1, the term 'outlier test' is misleading. Since the true population affiliation of an outlier candidate is not known, it would be more convenient to talk about 'tests to identify outlier candidates'. The term 'outlier candidate' is the more appropriate in hardly any practical context. For the sake of simplicity however, the term 'outlier' will be used in general for the remainder of this work. It will be clear from the context, if the true population affiliation is known or not.

3.1 Classification of Outlier Tests

There exists several types of outlier tests. Some tests only check a predefined number of suspicious extreme values. Other tests scan the whole dataset for outlying measurements without selecting suspicious candidates in advance. In the following sections, these concepts will be further explained and exemplary tests will be presented.

3.1.1 Tests for a Fixed Number of Outlier Candidates

In many practical applications, the user identifies one or a few suspicious values within the dataset based on his subjective impression and his experience in the field. Hence, he wishes to test a fixed number of predefined outlier candidates. A corresponding outlier test for one suspicious value will be based on the following informal hypotheses:

> H_0 : The suspicious value is no true outlier and thus belongs to P_{int},
>
> versus
>
> H_1 : The suspicious value really is an outliers and belongs to P_{cont}.

Equivalent hypotheses will be formulated if several predefined outlier candidates are to be tested:

> H_0 : At least one of the suspicious values belongs to P_{int},
>
> versus
>
> H_1 : All suspicious values belong to P_{cont}.

Note that an outlier test based on the above hypotheses do not provide a global answer with regard to the presence or absence of any outliers. Therefore, this kind of test is only appropriate if the data analyst who decides which values seem suspicious is well experienced with the type of data situation.

Many stepwise procedures for the identification of outliers have been proposed in this context, compare for example [Hawkins, 1980] (Chapter 5, Pages 63-66). An exemplary outlier test for one predefined extreme value is the well known Grubb's Test [Grubbs, 1950]. Here, the absolute difference between the mean value of data and the outlier candidate divided by the standard deviation is compared to predefined distributional quantiles. Other examples can be found in [Hawkins, 1980] (Chapter 3, Pages 27-40 and Chapter 5, Pages 52-67).

3.1.2 Tests to Check the Whole Dataset

As it has been explored in Section 2.3.2, it is not always obvious which values are extreme realizations with respect to the underlying statistical distribution or model. Therefore outlier candidates may be hard to distinguish. Hence, an outlier test is needed which scans the whole dataset for the presence of any outlier. In this case, hypotheses will be stated as follows:

$$H_0 : \text{The dataset does not contain any outlier,}$$
versus \hfill (3.1.1)
$$H_1 : \text{There are outliers present in the dataset.}$$

Usually, the outlier test should not only be able to state the presence of outliers but to identify them, as well. Therefore, most global outlier test are constructed by calculating predefined outlier limits, which are given in form of confidence limits for the specific comparison measure. Test decision is made by comparing the measure of interest to the particular outlier limits. H_0 is rejected if any of the measurement values exceeds the given outlier limits. All measurement values which lay outside the predefined outlier limits are identified as outlier. For a dataset of sample size n, the global test (3.1.1) is thus given as a multiple test situation consisting of n single tests. For $i = 1, ..., n$ the hypotheses for these single tests are given by:

$$H_{i,0} : \text{The } i^{\text{th}} \text{ measurement value is no outlier,}$$
versus \hfill (3.1.2)
$$H_{i,1} : \text{The } i^{\text{th}} \text{ measurement value is an outlier.}$$

As (3.1.1) is a multiple test situation, this will lead to the accumulation of first order errors. Therefore, the local significance levels α_{loc} for the single tests (3.1.2) should be adjusted in order to keep a global significance level α_{glob}.

The most common method to adjust the local significance levels is the well known Bonferroni adjustment, compare [Hsu, 1996] (Chapter 1, Page 13):

$$\alpha_{\text{loc}} = \frac{\alpha_{\text{glob}}}{n}.$$

The method of Bonferroni is the simplest and most flexible adjustment procedure proposed in the literature. It can be used in any multiple testing situation, requires no further statistical assumptions and is simple and fast to calculate. Unfortunately, it may lead to a notable loss of power - especially for a high number of strongly correlated tests. Therefore, an outlier test based on the Bonferroni adjustment should always be accompanied by a visual inspection of the data.

A less conservative alternative which however requires more computational effort is the stepwise

Bonferroni-Holmes procedure proposed by [Holm, 1979]. However, there exist many other methods to adjust the local significance levels in a multiple testing situation. An overview of the different procedures is given in [Hochberg, Tamhane, 1987] and [Hsu, 1996]. The adjustment procedures differ with respect to the power loss, the underlying computational effort and the required statistical assumptions.

As the focus of the outlier test lays not exclusively on the global test hypotheses (3.1.1) but also on the local test hypotheses (3.1.2), this will lead to extended performance measures for the statistical test. For the test (3.1.1), the power is defined as the probability to detect outliers under the condition that the dataset truly contains outliers. This does not imply that the test identifies the right observations as outliers. The probability to identify the right observations as outliers under the condition that the dataset contains outliers will be a supplementary measure of performance. These performance measures are discussed by [Hawkins, 2002] (Chapter 2, Pages 13-14).

The LORELIA Residual Test developed in the context of this work is an example for an outlier test which scans the whole dataset for outlying measurements. Other examples can be found in [Davies, Gather, 1993] and [Hawkins, 2002] (Chapter 5, Pages 57-63).

3.2 Formulation of the Test Hypotheses

The formulation of the test hypotheses is highly related to the predefined goal of outlier identification which allows to classify them in different subgroups.

3.2.1 Discordancy Tests

If the goal of outlier identification is to eliminate the existing outliers, the task is to separate the contaminating population P_{cont} from the population of interest P_{int} and to exclude outliers before any further data analysis. Test hypotheses for a corresponding outlier test will be formulated as follows:

H_0 : All observations fit with the given statistical assumptions for P_{int},

versus

H_1 : There exist observations which are discordant to the given statistical assumption of P_{int}.

Tests with the above hypotheses will be referred as 'discordancy tests' as stated in [Barnett, Lewis, 1994] (Chapter 2, Pages 37-38).

3.2.2 Incorporation of Outliers

If there exist observations which do not fit with the stated model or distributional assumptions for P_{int}, it may be appropriate not to eliminate these values but to explain them by supplementary or new assumptions. In [Barnett, Lewis, 1994] (Chapter 2, Page 39) this is referred as 'incorporation of outliers'. There exist several ways to incorporate outliers which will be presented in the following sections.

3.2.2.1 The Inherent Hypotheses

As it has been explained in Section 2.4.2, extremeness of measurement values may be due to wrong model assumptions for the population of interest. Test hypotheses should thus state an alternative model or distribution for the whole dataset:

---

versus

H_0 : All data are explained well by the given model or distribution,

H_1 : All data are explained better by another predefined model or distribution.

Since test based on these hypotheses are based on the assumption that the whole dataset belongs to the same population, they are referred as 'inherent hypotheses' in [Barnett, Lewis, 1994] (Chapter 2, Page 46). The alternative model or distribution stated in H_1 may differ only by a change of the parameters but it may as well be a completely different model or distribution.

3.2.2.2 The Deterministic Hypotheses

Instead of adjusting the statistical assumptions for the whole dataset, hypotheses may state an alternative model or distribution for the suspicious values only:

> versus
>
> H_0 : All data are explained well by the given model or distribution,
>
> H_1 : Some suspicious values are explained better by another predefined model or distribution.

In [Barnett, Lewis, 1994] (Chapter 2, Page 45), these hypotheses are called 'deterministic'. The deterministic alternative is closely related to the 'mixed model alternative' which will be presented in the following.

3.2.2.3 The Mixed Model Alternative

Mixed models and distributions have been defined in Definition 2.2 in Section 2.3.1. In the case of existing extreme observations, an alternative mixed model or distribution is stated, which explains the outlying values as well as the normal data. Hypotheses are given as follows:

> versus
>
> H_0 : All data are explained well by the given model or distribution,
>
> H_1 : Most observations are well explained by the given assumptions but with a small probability $1 - p$ the observations follow another model or distribution.

The problem here is to estimate the distribution parameters for the mixed model as parameter estimates for the contaminating distribution or model are usually based on very few data points.

3.3 Problems and Test Limitations

In this Section some problems which may lead to incorrect outlier classifications are pointed out.

3.3.1 The Masking Effect

The presence of several outliers in a dataset may avoid the identification of even one outlier. This is called the 'masking effect'. To illustrate this, consider a common outlier test in which the outlier candidate is compared to its right or left neighbor respectively. A big difference indicate that the outlier candidate is isolated and hence really is a true outlier. A small difference is expected to indicate that the measurement value is not isolated. However, if several outliers lay close together, this may lead to a masking effect.

In the following graphic, two data situations are presented. In the first data situation, one outlier is identified since the observation is isolated from all other data points. In the second data situation, one supplementary extreme value is included in the dataset so there are two outliers present which lay close together. The masking effect now avoids that the outlier is correctly identified.

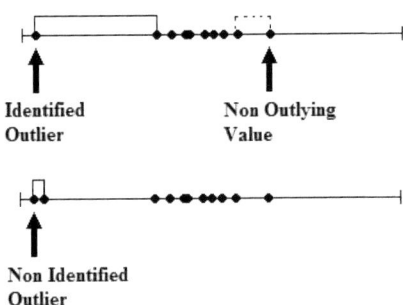

Figure 3.1: The Masking Effect

Examples for outlier tests suffering from the masking effect are given by [Davies, Gather, 1993], [Acuña, Rodriguez, 2005] and [Burke, 1999].

3.3.2 The Swamping Effect

Whereas a masking effect avoids the identification of true outliers in case of several existing outliers, the swamping effect causes the identification of too many outliers. Outlier tests which test a predefined fixed number of outlier candidates may suffer under such a swamping effect. For example, consider an outlier test which compares the mean of the two most extreme values to their next neighbor:

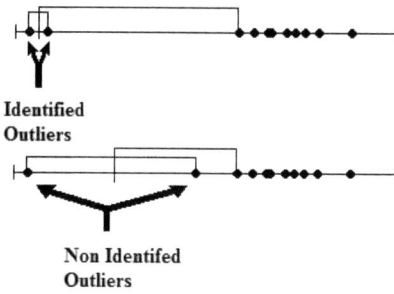

Figure 3.2: The Swamping Effect

In the first data situation, the two outliers are correctly identified since their mean is far away from the main body of the data. In the second data situation, only one true outlier exists. The mean between the outlier and its next neighbor however is still very large compared to the values of the remaining dataset. Thus, both values will be classified as outliers by this test.

Examples for the swamping effect are discussed in [Davies, Gather, 1993] and [Acuña, Rodriguez, 2005].

3.3.3 The Leverage Effect

The estimation of non robust linear regression parameters can be influenced substantially by so called 'leverage points'. Leverage points are measurement values at the edge of the measuring range which are isolated to the main body of the data. Varying values of leverage points lead to very different parameter estimates and thus influence the identification of outliers.

To illustrate this, consider the following data table containing two options for the last observation 15. In both cases, observation 15 is isolated from the main body of the data and is thus a leverage point.

Obs	x	y
1	1	8
2	1.5	10
3	2	11.5
4	1	7.9
5	2.1	11.2
6	0.9	7.9
7	1.1	8.4
8	1.4	9.1
9	1.3	9
10	1.6	9.8
11	1.7	10
12	1.7	9.9
13	1.8	10.5
14	1.9	10.7
15	4	22 / 17.1

Table 3.1: One Dataset with two Different Leverage Points

A simple linear regression fit for the first dataset delivers the following results:

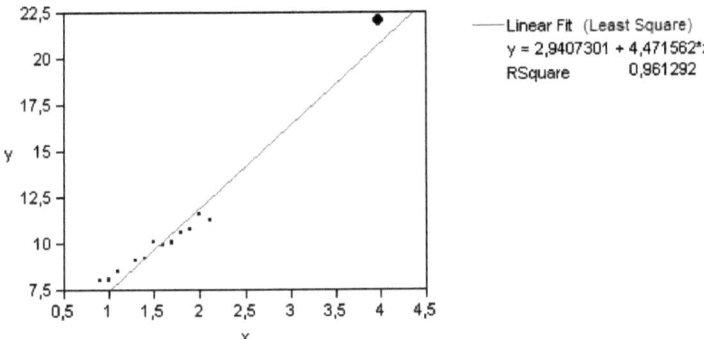

Figure 3.3: Linear Regression with the First Leverage Point Included

With $R^2 = 0.961292$ the linear model seems highly appropriate for the given dataset. Now, consider the linear regression fit with the second leverage point included:

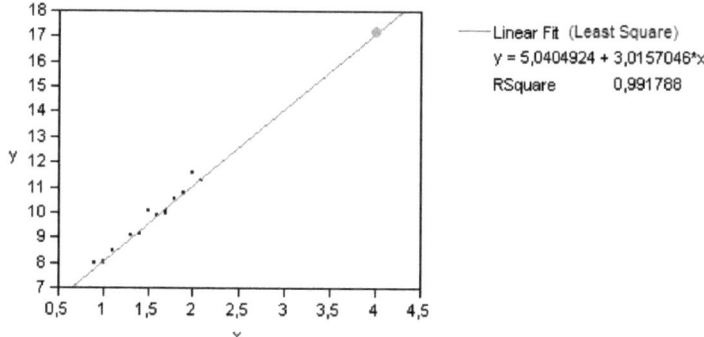

Figure 3.4: Linear Regression with the Second Leverage Point Included

With $R^2 = 0.991788$ the model adjustment has improved a lot. The parameter estimates are very different from those of the first dataset. The influence of the leverage point is obvious.

In practical applications, parameter estimates which may suffer from a leverage effect must always be handled and interpreted with care. Leverage points may bias the estimates but they can as well stabilize them. Therefore, a supplementary data analysis without the leverage point may be helpful. Thus, consider the regression fit with both leverage points excluded:

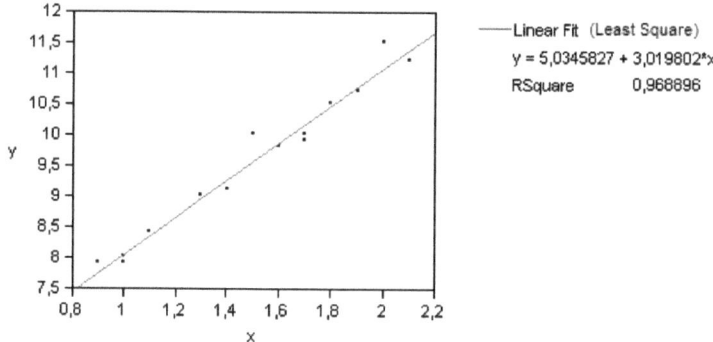

Figure 3.5: Linear Regression without the Leverage Points

Here $R^2 = 0.968896$, so the model adjustment is superior to the one with the first leverage point included, but inferior to the one with the second leverage point included. Moreover, the parameter estimates are very similar to those for the second dataset. Hence, the second leverage point stabilize the parameter estimates, whereas the first bias them. A discussion on leverage points and how to deal with them is given in [Rousseeuw, Zomeren, 1990].

Chapter 4

Evaluation of Method Comparison Studies

Method comparison studies are performed to evaluate the relationship between two measurement series. In clinical chemistry, they may for example be conducted to compare two measurement methods, two instruments or two diagnostic tests. Often the aim is to compare the performance of a newly developed method to a well established reference method. Several samples at different concentration levels are measured with both methods or instruments, respectively. These measurement tuple series are compared in order to show the equivalence between the two methods or to detect systematic differences.

There exist several possibilities to evaluate method comparison studies. A common approach, which is presented in Section 4.1, is to calculate the differences between two corresponding measurement values and to analyze these differences. Another possibility to compare two measurement series, which is discussed in Section 4.2, is the fit of a linear regression line. Both approaches require special distributional assumptions, which are not always met for the original data, but may be fulfilled after an appropriate data transformation, for example a log transformation or a generalized log transformation, compare [Rocke, Lorenzato, 1995]. A general overview of the different evaluation procedures and possible data transformations is given by [Hawkins, 2002]. The different measurement error models used in this context are summarized in the work of [Cheng, Ness, 1999].

4.1 Comparison by the Method Differences

In order to compare two measurement series, it is a common practice to determine the differences between the corresponding x- and y-values and to compare its average and standard deviation to some predefined limits to test equivalence. There exists several alternatives to calculate the differences.

4.1.1 The Absolute Differences

One common procedure is discussed by [Altman, Bland, 1983], [Bland, Altman, 1986], [Bland, Altman, 1995] and [Bland, Altman, 1999] which propose to use the absolute differences.

For $n \in \mathbb{N}$, let $x_1, ..., x_n$ and $y_1, ..., y_n$ be two measurement series corresponding to method M_x and M_y respectively. The observed measurement values are assumed to be described by:

$$x_i = c_i + \alpha_x + \epsilon_x \qquad (4.1.1)$$
$$y_i = c_i + \alpha_y + \epsilon_y, \text{ for } \alpha_x, \alpha_y \in \mathbb{R},\ i = 1, ..., n, \qquad (4.1.2)$$

where c_i is the unbiased, true concentration which is biased by the systematic additive term α_x respective α_y and the measurement error ϵ_x respective ϵ_y. The measurement errors ϵ_x and ϵ_y are realizations of the random variables:

$$E_x \sim N(0, \sigma_x^2), \qquad (4.1.3)$$
$$E_y \sim N(0, \sigma_y^2). \qquad (4.1.4)$$

The absolute differences, which are given by:

$$d_i^{\text{abs}} := y_i - x_i,\ i = 1, ..., n \qquad (4.1.5)$$

are therefore realizations of the random variable:

$$\begin{aligned} D_i^{\text{abs}} &= \alpha_y - \alpha_x + E_y - E_x \\ &\sim N(\alpha_y - \alpha_x, \sigma_x^2 + \sigma_y^2), \\ &=: N(\mu_{d_{\text{abs}}}, \sigma_{d_{\text{abs}}}^2). \end{aligned} \qquad (4.1.6)$$

Now, calculate the 97.5% confidence limits for the absolute differences D^{abs}:

$$\overline{d}^{\text{abs}} \pm z_{97.5\%} \cdot S_{d^{\text{abs}} d^{\text{abs}}}, \qquad (4.1.7)$$

where $z_{97.5\%}$ is the corresponding quantile of the normal distribution, $\overline{d}^{\text{abs}}$ is the mean and $S_{d^{\text{abs}} d^{\text{abs}}}$ is the empirical standard deviation of $d_1^{\text{abs}}, ..., d_n^{\text{abs}}$. In [Bland, Altman, 1986], these confidence limits are called the 'limits of agreement'. In order to test equivalence between the two methods, the limits of agreement are compared to predefined clinical reference values. Confidence bounds for the limits of agreement can be calculated as described by [Bland, Altman, 1986] and [Bland, Altman, 1999] in order to estimate the influence of the sampling error. Note that these limits of agreement are not robust against outliers since they are based on non robust location and scale estimators. Thus, the dataset should be checked for outliers in advance (compare Section 5.1).

Assumption (4.1.6) can be visually verified with the help of a scatter plot where the absolute differences d_i^{abs} are plotted against the means of the measurement values $\frac{x_i+y_i}{2}$.

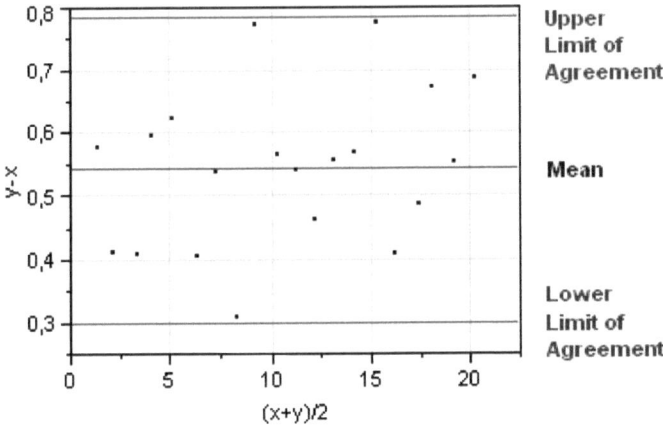

Figure 4.1: Method Comparison based on the Absolute Differences

The mean values $\frac{x_i+y_i}{2}$ are distributed as follows:

$$\frac{1}{2}(X_i + Y_i) \sim N\left(c_i + \frac{1}{2}(\alpha_x + \alpha_y), \frac{1}{4}(\sigma_x^2 + \sigma_y^2)\right), \text{ for } i = 1, ..., n. \qquad (4.1.8)$$

Thus, the mean values $\frac{x_i+y_i}{2}$ are only unbiased estimators for the true concentration c_i if:

$$\alpha_x = \alpha_y = 0. \qquad (4.1.9)$$

However, even if (4.1.9) is not fulfilled, the visual inspection of the scatter plot is still appropriate since a systematical bias on the horizontal axis will not affect the general normal assumption (4.1.6).

4.1.2 The Relative Differences

In many practical applications, the absolute differences will not have constant mean and variance over the measuring range. If the scatter plot reveals a proportional difference between measurement values, it will be more appropriate to consider a multiplicative random error.

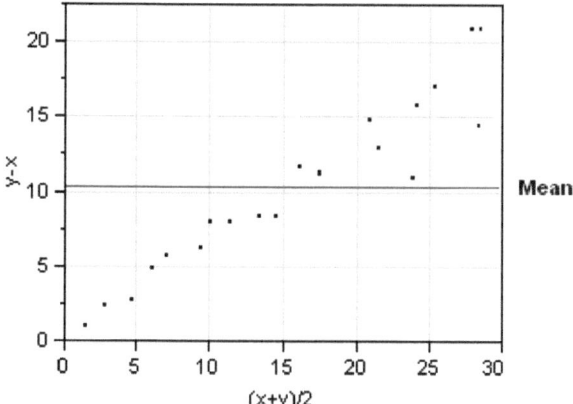

Figure 4.2: Proportional Bias Between Methods

The following model assumptions are considered:

$$x_i = c_i \cdot \beta_x + \epsilon_{x_i} \tag{4.1.10}$$
$$y_i = c_i \cdot \beta_y + \epsilon_{y_i}, \text{ for } \beta_x, \beta_y \in \mathbb{R}^+, i = 1, ..., n, \tag{4.1.11}$$

where the random errors are realizations of the random variables:

$$E_{x_i} \sim N(0, c_i^2 \cdot \sigma_x^2), \tag{4.1.12}$$
$$E_{y_i} \sim N(0, c_i^2 \cdot \sigma_y^2), \text{ for } i = 1, ..., n. \tag{4.1.13}$$

Note that the error variances in (4.1.12) and (4.1.13) depend on the true concentrations c_i.

The absolute differences are thus realizations of:

$$\begin{aligned} D_i^{\text{abs}} &= c_i \cdot (\beta_y - \beta_x) + E_{y_i} - E_{x_i} \\ &\sim c_i \cdot N(\beta_y - \beta_x, \sigma_x^2 + \sigma_y^2), \text{ for } i = 1, ..., n. \end{aligned} \tag{4.1.14}$$

The variance and the mean of the absolute differences are not constant here, but increase proportionally in c_i. By (4.1.14), this corresponds to the assumption of a constant coefficient of variance for th random errors. Note that the true concentrations c_i are not known here! They therefore have to be estimated, which is done by the mean of the observed measurement values:

$$\widehat{c}_i = \frac{y_i + x_i}{2}, \text{ for } i = 1, ..., n. \tag{4.1.15}$$

The mean values are distributed as follows:

$$\frac{1}{2}(X_i + Y_i) \sim N\left(\frac{c_i}{2}(\beta_x + \beta_y), \frac{c_i^2}{4}(\sigma_x^2 + \sigma_y^2)\right), \text{ for } i = 1, ..., n. \tag{4.1.16}$$

Again, the mean values $\frac{x_i+y_i}{2}$ are only unbiased estimators for the true concentration c_i if:

$$\beta_x = \beta_y = 1. \tag{4.1.17}$$

The normalized relative differences are defined by:

$$d_i^{\text{normrel}} := \frac{y_i - x_i}{\frac{1}{2} \cdot (y_i + x_i)}, \text{ for } i = 1, ..., n. \tag{4.1.18}$$

If the mean $\frac{x_i+y_i}{2}$ is chosen as an estimate for the true concentration c_i, they are approximately distributed as:

$$D_i^{\text{normrel}} \overset{\text{approx}}{\sim} \frac{2}{(\beta_x + \beta_y)} \cdot N(\beta_y - \beta_x, \sigma_x^2 + \sigma_y^2),$$
$$=: N(\mu_{d_{\text{normrel}}}, \sigma_{d_{\text{normrel}}}^2). \tag{4.1.19}$$

The normalized relative differences have constant mean and variance. Hence the limits of agreement can be calculated as the 97.5% confidence limits:

$$\overline{d}^{\text{normrel}} \pm 1.96 \cdot S_{d^{\text{normrel}}d^{\text{normrel}}}, \tag{4.1.20}$$

where $\overline{d}^{\text{normrel}}$ is the mean and $S_{d^{\text{normrel}}d^{\text{normrel}}}$ is the empirical standard deviation of $d_1^{\text{normrel}}, ..., d_n^{\text{normrel}}$.

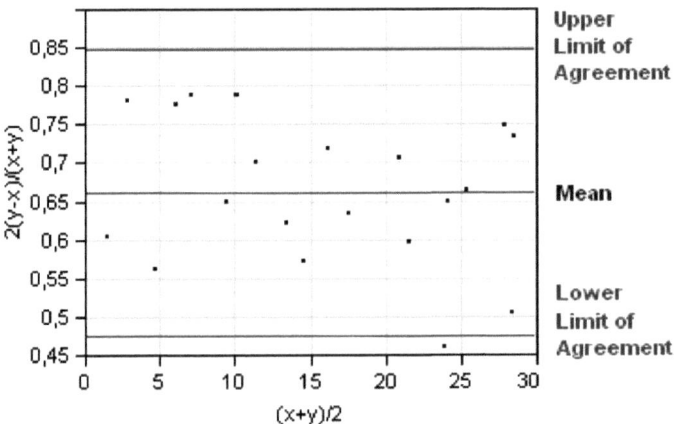

Figure 4.3: Method Comparison based on the Normalized Relative Differences

The dataset for the above scatter plot is the same as in Figure 4.2. Whereas Figure 4.2 clearly shows that the absolute differences are not normally distributed, the normal assumptions seems appropriate for the normalized relative differences plotted in Figure 4.3.

In the literature, the special case that one method (here M_x) is free of random error is often considered, as well. This correspond to the following model assumptions:

$$x_i = c_i \tag{4.1.21}$$
$$y_i = c_i \cdot \beta_y + \epsilon_{y_i}, \text{ for } \beta_y \in \mathbb{R}^+, i = 1, ..., n, \tag{4.1.22}$$

with ϵ_y being a realization of the random variable:

$$E_{y_i} \sim N(0, c_i^2 \cdot \sigma_y^2), \text{ for } i = 1, ..., n. \tag{4.1.23}$$

In this case, the absolute differences have the following distribution:

$$\begin{aligned} D_i^{\text{abs}} &= c_i \cdot (\beta_y - 1) + E_{y_i} \\ &\sim c_i \cdot N(\beta_y - 1, \sigma_y^2), \text{ for } i = 1, ..., n. \end{aligned} \tag{4.1.24}$$

Since method M_x is free of random error, the true concentration c_i is known here and does not have to be estimated. Therefore, the relative differences can be considered:

$$d_i^{\text{rel}} := \frac{y_i - x_i}{x_i}, \text{ for } i = 1, ..., n, \tag{4.1.25}$$

which are realizations of:

$$\begin{aligned} D^{\text{rel}} &\sim N(\beta_y - 1, \sigma_y^2), \\ &=: N(\mu_{d_{\text{rel}}}, \sigma_{d_{\text{rel}}}^2). \end{aligned} \tag{4.1.26}$$

Again the limits of agreement can be calculated as the 97.5% confidence limits:

$$\overline{d}^{\text{rel}} \pm 1.96 \cdot S_{d^{\text{rel}}d^{\text{rel}}}, \tag{4.1.27}$$

where $\overline{d}^{\text{rel}}$ is the mean and $S_{d^{\text{rel}}d^{\text{rel}}}$ is the empirical standard deviation of $d_1^{\text{rel}}, ..., d_n^{\text{rel}}$. Assumption (4.1.26) can be verified by plotting x_i against the relative differences $\frac{y_i - x_i}{x_i}$.

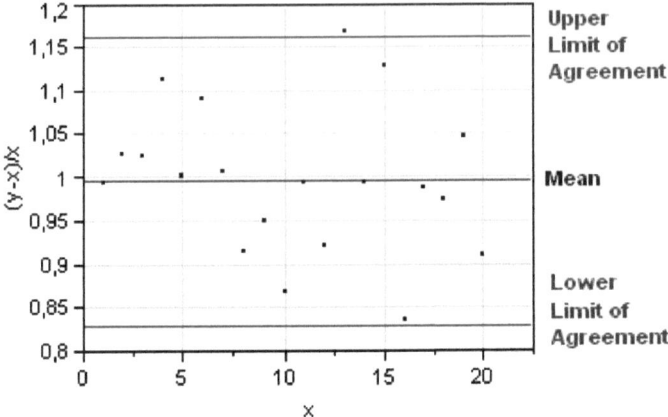

Figure 4.4: Method Comparison based on the Relative Differences

4.2 Comparison with Regression Analysis

The statistical comparison of two measurement series is often evaluated by fitting a linear regression line. The outcomes of the two methods which are to be compared are plotted against each other and a regression line is calculated. The evaluation of method comparison studies by regression analysis is discussed in [Hartmann et al. 1996].

Consider the following model assumptions as described by [Fuller, 1987] (Chapter I, Page 1):

$$x_i = \underbrace{\alpha_x + \beta_x \cdot c_i}_{=:\tilde{x}_i} + \epsilon_{x_i} \qquad (4.2.1)$$

$$y_i = \underbrace{\alpha_y + \beta_y \cdot c_i}_{=:\tilde{y}_i} + \epsilon_{y_i}, \text{ for } i = 1, ..., n, \qquad (4.2.2)$$

where c_i is the true concentration and $\tilde{x}_i$, $\tilde{y}_i$ are the expected measurement values for method M_x and M_y respectively which are exposed to the measurement errors ϵ_{x_i} and ϵ_{y_i}. Without loss of generality it will be assumed that:

$$x_i = c_i + \epsilon_{x_i} \qquad (4.2.3)$$

$$y_i = \alpha + \beta \cdot c_i + \epsilon_{y_i}, \text{ for } i = 1, ..., n. \qquad (4.2.4)$$

The measurement errors ϵ_{x_i} and ϵ_{y_i} are assumed to be realizations of the random variables:

$$E_{x_i} \sim N(0, \sigma_{x_i}^2), \quad (4.2.5)$$

$$E_{y_i} \sim N(0, \sigma_{y_i}^2), \text{ for } i = 1, ..., n. \quad (4.2.6)$$

The observed measurement values are hence realizations of the random variables:

$$X_i = \widetilde{x}_i + E_{x_i} \sim N(\widetilde{x}_i, \sigma_{x_i}^2) \quad (4.2.7)$$

$$Y_i = \widetilde{y}_i + E_{y_i} \sim N(\widetilde{y}_i, \sigma_{y_i}^2), \text{ for } i = 1, ..., n. \quad (4.2.8)$$

By (4.2.1) to (4.2.4), a linear relationship between the expected measurement values is assumed:

$$\widetilde{y}_i = \alpha + \beta \cdot c_i = \alpha + \beta \cdot \widetilde{x}_i, \quad \text{for } \alpha \in \mathbb{R}, \ \beta \in \mathbb{R} \setminus \{0\}, \ i = 1, ..., n. \quad (4.2.9)$$

Since the expected measurement values are not known, α and β have to be estimated by regression procedures.

For equivalent methods M_x and M_y the parameter estimates will be given by:

$$\widehat{\beta} \approx 1,$$
$$\widehat{\alpha} \approx 0.$$

A proportional bias between the two methods will be given if

$$\widehat{\beta} \neq 1.$$

Note that by assumptions (4.2.1) and (4.2.2) the regression method has to take random errors in both axes into account. Therefore ordinary least square regression is not appropriate in this context.

There exists a variety of robust and non robust regression methods. Since outlying measurements can effect the estimates of slope and intercept for non robust regression, the recommendation is to use robust regression procedures. Common robust regression methods will be presented and discussed in the following section.

4.2.1 Robust Regression Methods

There exist a variety of robust regression methods which are based on different statistical assumptions. An overview can be found in [Rousseeuw, Leroy, 1987]. The procedure recommended in this work is Passing-Bablok regression, which will be presented in the following. In the literature, principal component analysis and standardized principal component analysis are often referred as robust procedures as well, although they are parametric. Both methods are special cases of the more general Deming regression described in [Deming, 1943] and [Linnet, 1998]. Deming regression is the most commonly used procedure in the context of method comparison studies and will therefore be presented in this section as well, although it can not be regarded as a robust procedure. A comparison and a detailed discussion of the above regression methods can be found in [Stökl et al., 1998]. Other robust regression methods are proposed by [Brown, 1988], [Feldmann, 1992], [Ukkelberg, Borgen, 1993], [Hartmann et al. 1997] and [Olive, 2005]. The following sections are basically referred to the work of [Haeckel,1993] (Chapter 11, Pages 212-226).

4.2.1.1 Deming Regression

For the Deming regression, the measurement errors ϵ_{x_i} and ϵ_{y_i} are assumed to be realizations of the random variables:

$$E_{x_i} \sim N(0, \sigma_x^2), \qquad (4.2.10)$$

$$E_{y_i} \sim N(0, \sigma_y^2), \text{ for } i = 1, ..., n. \qquad (4.2.11)$$

Note, that the error variances are assumed to remain constant over the measuring range here. Further, a known ratio of error variances is assumed:

$$\frac{\sigma_y^2}{\sigma_x^2} = \eta^2, \text{ for a known } \eta \in \mathbb{R}^+ \setminus \{0\}. \qquad (4.2.12)$$

Deming regression minimizes the squared skew residuals, where the residual slope is given by $-\eta$. The minimization of the skew residuals is equivalent to the minimization of the orthogonal residuals after a respective transformation of the y-values:

$$y_i^t := \frac{y_i}{\eta}, \; i = 1, ..., n. \qquad (4.2.13)$$

After transformation, Deming regression thus corresponds to the common orthogonal least square regression or principal component analysis.

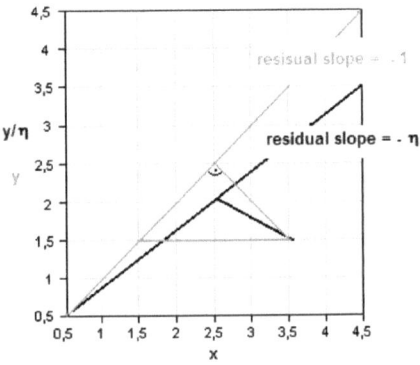

Figure 4.5: The Concept of Deming Regression

In the above plot, the black line corresponds to the regression line for the original measurement values with skew residuals. The gray line shows the regression line after the transformation of the y-values. The residuals are orthogonal now.

4.2.1.2 Principal Component Analysis (PCA)

Regression analysis by principal component decomposition is equivalent to orthogonal least square regression. It is a special case of the more general Deming regression described in the previous section. For a multivariate dataset the principal components are chosen iteratively. The first principal component is the vector with the direction of the highest variance in data dispersion. The other principal components are chosen in the same way with the restriction to be orthogonal to each other. For a two dimensional dataset, this is equivalent to orthogonal least square regression. PCA is based on the assumption, that the measurement errors are distributed as follows:

$$E_{x_i}, E_{y_i} \overset{iid}{\sim} N(0, \sigma^2), \text{ for } i = 1, ..., n, \qquad (4.2.14)$$

which correspond to the special case of equal error variances in (4.2.5) and (4.2.6) or an error variance ration of $\eta = 1$ in (4.2.12).

The slope estimator for principal component analysis is given as:

$$\widehat{\beta}_{\text{PCA}} := \frac{S_{yy}^2 + S_{xx}^2 + \sqrt{\left(S_{yy}^2 - S_{xx}^2\right)^2 + 4 \cdot S_{yx}^2}}{2 \cdot S_{yx}}, \qquad (4.2.15)$$

where S_{xx} and S_{yy} are the empirical standard deviations of the x- and y-values, respectively and S_{yx}^2 is the corresponding empirical covariance. The estimator for the intercept is defined through:

$$\widehat{\alpha}_{\text{PCA}} := \overline{y} - \widehat{\beta}_{\text{PCA}} \cdot \overline{x}. \qquad (4.2.16)$$

As the orthogonal residuals are always smaller or equal to the vertical residuals, outlying measurements will intuitively influence the parameter estimates for orthogonal regression less than for ordinary least square regression. Nevertheless, the PCA can not be regarded as an outlier resistant regression method since it is still based on non robust parameter estimators.

4.2.1.3 Standardized Principal Component Analysis (SPCA)

Standardized principal component analysis is a special case of Deming regression, as well. Here, the error variance ratio is assumed to be given by:

$$\frac{\sigma_y^2}{\sigma_x^2} = \beta^2, \qquad (4.2.17)$$

where β is the true slope given in (4.2.4). The above functional relationship between the true slope β and the error variances is assumed in order to reduce the parameters which has to be estimated. By (4.2.17), the random errors are realizations of:

$$E_{x_i} \overset{iid}{\sim} N(0, \sigma_x^2), \qquad (4.2.18)$$

$$E_{y_i} \overset{iid}{\sim} N(0, \underbrace{\beta^2 \cdot \sigma_x^2}_{=:\sigma_y^2}), \text{ for } i = 1, ..., n. \qquad (4.2.19)$$

Note that (4.2.19) correspond to the assumption that:

$$y_i = \alpha + \beta \cdot c_i + N(0, \beta^2 \cdot \sigma_x^2), \text{ for } i = 1, ..., n. \qquad (4.2.20)$$

For $\alpha = 0$ this is equivalent to the error model for the (normalized) relative differences given in (4.1.11) and (4.1.22).

If $\widehat{\beta}_{\text{SPCA}}$ is the slope of the first principal component, the second component in SPCA will correspond to a slope of $-\widehat{\beta}_{\text{SPCA}}$. Note that the slope of the second component determines the slope of the residuals, as well. The regression parameter estimators are given as:

$$\widehat{\beta}_{\text{SPCA}} := \text{sign}(S_{yx}) \cdot \sqrt{\frac{S_{yy}^2}{S_{xx}^2}} \qquad (4.2.21)$$

and

$$\widehat{\alpha}_{\text{PCA}} := \overline{y} - \widehat{\beta}_{\text{PCA}} \cdot \overline{x}. \qquad (4.2.22)$$

Note, that the use of SPCA is only appropriate if assumption (4.2.20) is fulfilled. A proportional bias between methods can be visually detected with the help of a scatter plot as given in Figure 4.2.

A geometrical interpretation of PCA and SPCA is given in Figure 4.6:

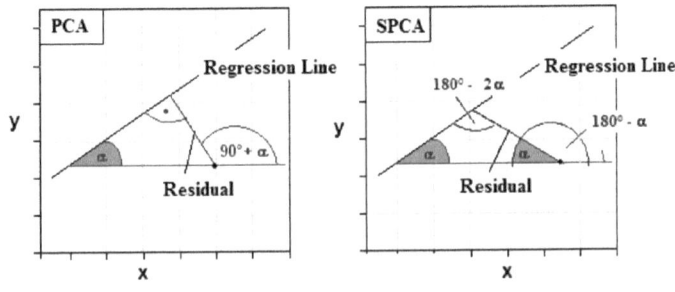

Figure 4.6: Residuals for PCA and SPCA

4.2.1.4 Passing-Bablok Regression

Passing-Bablok regression requires the least strong statistical assumptions among the presented regression procedures. The following model assumptions have to be fulfilled:

- The random errors E_{x_i} and E_{y_i} of method M_x respective M_y come from the same type of an arbitrary continuous distribution. Note that this is a much weaker assumption than stated in (4.2.5) and (4.2.6).

- The error variances may depend on the true concentration c_i, so they are not required to be constant over the measuring range, as required for Deming regression, but they have to remain proportional:

$$\frac{\sigma_{y_i}^2}{\sigma_{x_i}^2} = \eta^2, \text{ for } \eta \in \mathbb{R}^+ \setminus \{0\}, \ i = 1, ..., n. \quad (4.2.23)$$

Note that the parameter η does not have to be known here!

- The true slope defined by (4.2.9) is close to 1:

$$\beta \approx 1. \quad (4.2.24)$$

Passing-Bablok regression as described in [Passing, Bablok, 1983], [Passing, Bablok, 1984] and [Bablok et al. 1988] is based on the concept of Theil's regression [Theil 1950]:

In a first step, the slopes of the straight lines between all possible data pairs are calculated:

$$S_{ij} := \begin{cases} \frac{y_i - y_j}{x_i - x_j}, & \text{for } x_i \neq x_j \text{ and } y_i \neq y_j, \\ \infty, & \text{for } x_i = x_j \text{ and } y_i < y_j, \\ -\infty, & \text{for } x_i = x_j \text{ and } y_i > y_j, \\ 0, & \text{for } y_i = y_j, \end{cases} \text{ for all } i, j = 1, ..., n.$$

Without loss of generality, it will be assumed that:

$$|S_{ij}| \in \mathbb{R} \setminus \{0\}, \text{ for all } i, j = 1, ..., n.$$

This is appropriate, since (X, Y) is a continuous bivariate random variable and hence the probability to observe values with $S_{ij} = 0$ or $|S_{ij}| = \infty$ equals 0. Now, consider the ranked sequence $S_{(1)} \leq S_{(2)} \leq ... \leq S_{(n)}$.

In [Theil 1950], the median of the above sequence is defined as the slope estimator. Note however, that the S_{ij} are not statistically independent. Hence, a simple median estimator may be biased. A bias correction is proposed in [Passing, Bablok, 1983] by the following offset:

$$K := \#\{S_{ij} : S_{ij} < -1, \text{ for } i, j = 1, ..., n\}. \quad (4.2.25)$$

The corrected median estimator is given as:

$$\widehat{\beta}_{\text{PB}} := \begin{cases} S_{\left(\frac{N+1}{2}+K\right)}, & \text{if } n \text{ is odd,} \\ \frac{1}{2} \cdot \left(S_{\left(\frac{N}{2}+K\right)} + S_{\left(\frac{N}{2}+1+K\right)} \right), & \text{if } n \text{ is even.} \end{cases} \qquad (4.2.26)$$

In ([Passing, Bablok, 1984], page 717) it is proved that this slope estimator is independent of the assignment of the methods to x and y. Moreover it is shown empirically that $\widehat{\beta}_{\text{PB}}$ is an unbiased slope estimator if the true relation between method M_x and M_y correspond to a slope of 1. The estimator for for the intercept is defined as:

$$\widehat{\alpha}_{\text{PB}} := \text{med}_{\{1 \leq i \leq n\}}(y_i - \widehat{\beta} \cdot x_i).$$

Since the error variances remain proportional over the measuring range by assumption (4.2.23) and the true slope β is assumed to be close to 1, it will be appropriate to consider the *orthogonal* residuals in order to describe the location of the data pairs (x_i, y_i) with respect to the Passing-Bablok regression line.

Note that as a non parametric method Passing-Bablok regression is more robust against outliers than PCA or SPCA.

Chapter 5

Common Outlier Tests for Method Comparison Studies

The presence of outliers in method comparison studies can influence statistical data analysis and hence may lead to wrong conclusions on equivalence or non equivalence for the two methods. This can basically be avoided by using robust statistical methods, which are resistant against the presence of a few outliers.

However, the identification of outliers is still a very important task if method comparison is based on robust statistical methods, since the presence of outliers can reveal valuable information on the measurement process. Outliers can indicate

- a lack of performance in one of the two methods. For example, a newly developed method may work equally good as the reference method for samples at the lower concentration range, but fails for higher concentrated samples.

- a problem with the specific sample. For example, a somehow contaminated sample may lead to unexpectedly large measurement values.

For the above reasons, identified outliers should always be carefully examined and reported.

Many different outlier test are described in the statistical literature. A general overview can be found in [Hawkins, 1980] and [Barnett, Lewis, 1994]. Most outlier tests are constructed to test a predefined number of extreme values, compare Section 3.1.1. From a visual inspection of the dataset however, it is not always obvious which values can be regarded as extreme. For example, typical datasets in method comparison studies often show a very inhomogeneous sample distribution - the main part of the data is accumulated at a low concentration range (sample results from the healthy part of the study population) whereas only some isolated values correspond to higher concentrated samples (sample results from the pathological part of the population). An outlier classification for isolated values is generally difficult, even if the corresponding values seem surprisingly extreme, as the local level of data evidence is very low. The other way round, extreme values may not be visually detected, if the graphical representation is inappropriate. Therefore, it is strongly recommended to scan the whole dataset for the presence of outliers, as described in Section 3.1.2. Unfortunately, the statistical literature offers very few suggestions how to deal with the special problem of a unknown number of outliers most of which require that the measurement values can be ranked with respect to their extremeness, compare for example [Hawkins, 1980] (Chapter 5, Pages 51-73). However, as a ranking for the extremeness of measurement values is often note possible as described above, these procedures are not appropriate in this context. Another intuitive approach which is followed by [Wadsworth, 1990] is to calculate predefined outlier limits, which are given in form of robust confidence limits for the specific comparison measure. This concept is closely related to the informal identification of outliers with the help of boxplots. Although the test proposed by [Wadsworth, 1990] is not very established in the statistical literature as it provides no proper type 1 error control, it is a simple solution to handle the special outlier problem described above and will therefore be used as the reference method in the context of this work.

5.1 Outlier Tests based on Method Differences

If method comparison is evaluated by using one of the difference measures presented in Section 4.1, outlier limits will be given as a confidence interval for the considered difference measure. The construction of those confidence intervals requires robust scale and location estimators. If the normal assumption (4.1.6), (4.1.19) respective (4.1.26) is met, it is recommended in [Wadsworth, 1990] (Chapter 16, Section 4) to use the median and the 68% median absolute deviation as sclae and location estimators. In the work of [Wadsworth, 1990] the following outlier limits are proposed:

$$\mathrm{med}(d^*) \pm 2.5 \cdot \mathrm{mad68}(d^*), \quad (5.1.1)$$

where d^* corresponds to the considered comparison measure, here d^{abs}, d^{rel} or d^{normrel}. The choice of the cutoff value 2.5 is not further explained in [Wadsworth, 1990]. The author refers this value to a low level of significance for the corresponding outlier test, which is not explicitly declared and points out that the choice of this value is to a certain extend arbitrary. The cutoff value should correspond to some high quantile of the the random variable:

$$\frac{D^* - \mathrm{med}(D^*)}{\mathrm{mad68}(D^*)}. \quad (5.1.2)$$

The construction of the outlier limits is similar to the definition of the limits of agreement, but they should not be confounded. The limits of agreement are used in order to test equivalence between the two methods when no outliers are present and are therefore based on non robust parametric estimators. The outlier limits proposed in (5.1.1) are much wider than the limits of agreement since they are used to identify surprisingly extreme measurements. They are robustly estimated to avoid masking effects. With the help of a scatter plot, outlier identification can now be done graphically:

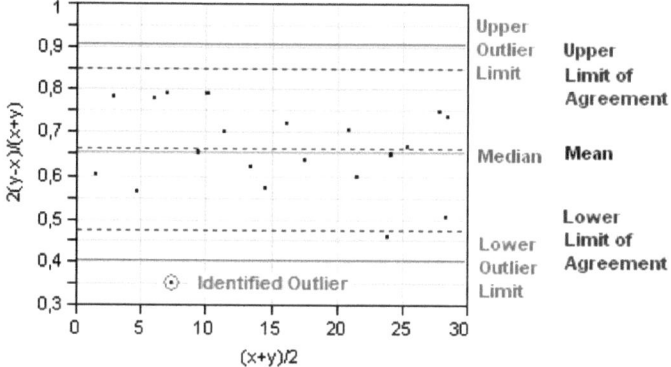

Figure 5.1: Outlier Identification Based on the Normalized Relative Differences

5.1.1 Problems and Limitations

The proposed outlier identification rule is not referred to a specific significance level, which is a clear drawback and complicates the comparison of its performance to other outlier tests. The multiple testing situation as described in Section 3.1.2 and the resulting problem of the accumulation of type 1 error rates is completely neglected here.

Moreover, the above outlier test is based on the strong statistical assumption that the considered difference measure (absolute, relative or normalized relative differences) are normally distributed with constant variances. However, the random error variances are often neither constant over the measuring range nor proportional to the true concentration, so none of the normal assumptions (4.1.6), (4.1.19) respective (4.1.26) is fulfilled. Although an appropriate data transformation may solve this problem, there exist many data situations in which standard transformations are useless.

5.2 Outlier Test based on Regression

There exist a variety of outlier tests for regression analysis in the statistical literature. However, most of them are based on ordinary least square regression. Outlier tests based on ordinary least square regression usually search for values with a high influence on the parameter estimates (leverage points) by comparing the parameter estimates with the suspicious values in- and excluded, compare [Rousseeuw, Leroy, 1987] (Chapter 6, Page 216). The most common method which is based on this approach is to calculate the external studentized residuals. But there exist a variety of other outlier tests which are based on these principles, compare for example [Rio et al., 2001] and [Xie, Wei, 2003].

In the context of method comparison studies however, leverage points do not necessarily correspond to outliers. On the contrary, as pointed out above, typical datasets in method comparison studies often show a very inhomogeneous sample distribution, so leverage points are rather a standard phenomenon. Moreover, if the dataset is supposed to contain outliers, it is highly recommended to use robust regression procedures, which will not be influenced much by the presence of outliers or leverage points. Therefore, most standard outlier tests to detect outliers in a linear model are not appropriate in this context.

If the evaluation of method comparison is based on a robust regression procedure, outliers correspond to measurement values with surprisingly high residuals. Therefore, outlier limits will be given in form of confidence limits for the considered residuals (orthogonal or skew). If the residuals are normally distribution and homoscedastic:

$$R_i \stackrel{iid}{\sim} N(0, \sigma_r^2), \text{ for } i = 1, ..., n, \tag{5.2.1}$$

the concepts of [Wadsworth, 1990] can be applied similar to (5.1.1) in Section 5.1, so outlier limits are defined as follows:

$$\text{med}(r) \pm 2.5 \cdot \text{mad68}(r). \tag{5.2.2}$$

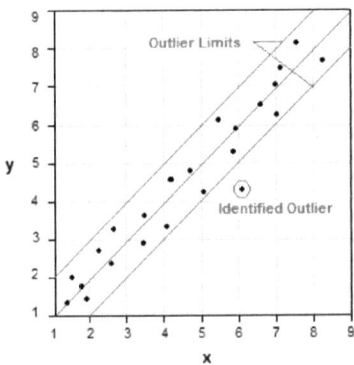

Figure 5.2: Confidence Bounds for the Residuals

5.2.1 Problems and Limitations

The outlier limits proposed by [Wadsworth, 1990] generally lead to the problems described in Section 5.1.1 independently of the choice of the underlying comparison measure. However, if the concepts of [Wadsworth, 1990] are applied to the residuals of a liner a regression model, the most serious problem is that assumption (5.2.1) is often not fulfilled for different reasons:

- If a parametric regression method is used which is based on minimizing the squared residuals (e.g. PCA or SPCA), assumption (5.2.1) will be violated since the residuals can not be regarded as *independent* realizations of the same random variable as they sum up to 0. For the non parametric Passing-Bablok regression however, the assumption of independently distributed residuals is not violated.

- In practical applications, the residual variances are often not constant over the measuring range. This heteroscedastic case is problematical, since the unknown true underlying residual variance model is often rather complex so standard transformations are useless. Appropriate model assumptions on the residual variance are difficult or even impossible to find.

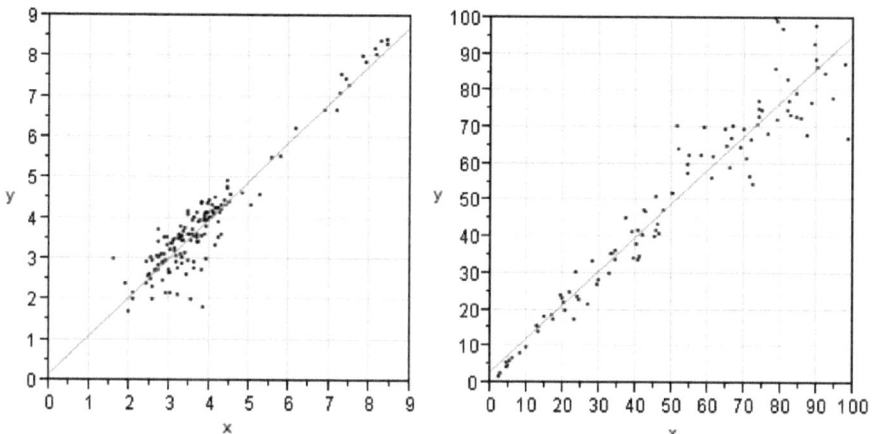

Figure 5.3: Examples for Heteroscedastic Residual Variance Models

Chapter 6

The New LORELIA Residual Test

As it has been explored in Chapter 5, most outlier tests for method comparison studies described in the literature are formulated exclusively for a fixed number of outlier candidates and require very strong statistical assumptions on the measurement error variance model or the residual variance model, respectively. If the true number of outliers is unknown and the underlying error variances are rather complex, these tests are therefore very limited in use. Also there exists a variety of different error models and respective data transformation rules (compare for example the models proposed by [Hawkins, 2002], [Cheng, Ness, 1999] and [Rocke, Lorenzato, 1995]), which allow to apply common outlier test like the test proposed by [Wadsworth, 1990] to a wider range of data situations, this is not a satisfying solution to the general problem. On the one hand, the choice of an appropriate data transformation rule is not always obvious and requires a good knowledge of the different error models described in the literature. Moreover this implies that every dataset should be handled differently as method comparison studies correspond to very heterogeneous data situations. It would be much more satisfying to suggest a general solution to the outlier problem independently of the underlying error variance model. On the other hand, there exist data situation in which none of the data transformations proposed in the literature are appropriate. So far, no outlier identification test has been proposed for these non standard data situations.

In this chapter a new outlier identification test for method comparison studies, which is based on Passing-Bablok regression will be presented. Passing-Bablok regression is chosen for the following reasons:

(i.) It is very robust against the presence of outliers,

(ii.) It takes random errors in both variables into account,

(iii.) The measurement error variances are not required to be constant.

However, Passing-Bablok regression may be replaced by another robust regression method if it fulfills the above requirements.

The new test is called the LORELIA Residual Test (LOcal RELIAbility Residual Test). The main concept of the new test is to construct robust, local confidence intervals for the orthogonal residuals to deal with the special problem of a heteroscedastic residual distribution for an unknown underlying error variance model. The new test is based on relaxed statistical assumptions in comparison to the test proposed by [Wadsworth, 1990] presented in Chapter 5 which will be explored in Section 6.1. In Section 6.2, the construction of the local outlier limits is deduced, which requires the definition of a local residual variance estimator. This estimator is calculated as a weighted sum of the squared observed residuals. In Section 6.3, the requirements for the construction of appropriate weights are discussed. The definition of the weighting function is given in Section 6.4. Finally in Section 6.5, the LORELIA Residual Test will be summarized and formally defined.

6.1 Statistical Assumptions for the New Test

Remember the model assumptions for Passing-Bablok regression presented in Section 4.2.1.4: The random errors E_{x_i} and E_{y_i} of method M_x and method M_y, respectively, are assumed to come from the same type of an arbitrary continuous distribution with proportional error variances, which are not necessarily required to remain constant over the measuring range. In this context, it will be assumed that the random errors both come from a normal distribution:

$$E_{x_i} \sim N(0, \sigma_{x_i}^2), \qquad (6.1.1)$$
$$E_{y_i} \sim N(0, \sigma_{y_i}^2), \text{ for } i = 1, ..., n, \qquad (6.1.2)$$

with equal error variances:

$$\frac{\sigma_{y_i}^2}{\sigma_{x_i}^2} = 1, \; i = 1, ..., n. \qquad (6.1.3)$$

As the error variance ratio equals 1, it is appropriate to consider the orthogonal residuals to describe the location of measurement values with respect to the regression line. the orthogonal residuals for Passing-Bablok regression are calculated as:

$$r_i = \frac{y_i - \widehat{\alpha}_{PB} - \widehat{\beta}_{PB} \cdot x_i}{\sqrt{1 + \widehat{\beta}^2}}, \text{ for } i = 1, ..., n. \qquad (6.1.4)$$

To deduce the distributional properties of the orthogonal residuals R_i consider Figure 6.1. Remember that by (4.2.9) in Section 4.2, a linear relationship between the expected measurement values $\widetilde{y}_i = \alpha + \beta \widetilde{x}_i$ is assumed for every $i = 1, ..., n$. The observed measurement values x_i and y_i differ from the expected measurement values by the measurement errors ϵ_{x_i} and ϵ_{y_i}, respectively.

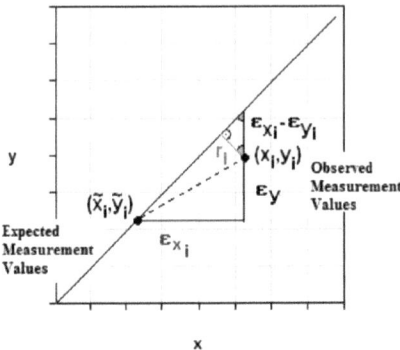

Figure 6.1: The Orthogonal Residuals

By the Pythagorean Theorem the orthogonal residuals are realizations of:

$$R_i := \frac{1}{\sqrt{2}} \left(E_{x_i} - E_{y_i} \right), \quad \text{for} \quad i = 1, ..., n.$$

Because of (6.1.1) and (6.1.2) it holds that:

$$(E_{x_i} - E_{y_i}) \sim N(0, 2 \cdot \sigma_{x_i}^2), \quad \text{for} \quad i = 1, ..., n.$$

Hence, the residual distribution is given as:

$$R_i \sim \frac{1}{\sqrt{2}} N \left(0, 2 \cdot \sigma_{x_i}^2 \right) = N(0, \sigma_{x_i}^2) := N(0, \sigma_{r_i}^2), \quad \text{for} \quad i = 1, ..., n. \qquad (6.1.5)$$

The residual distribution thus correspond to the distribution of the random errors under the assumption that the Passing-Bablok estimators $\widehat{\alpha}_{PB}$ and $\widehat{\beta}_{PB}$ are unbiased estimates of the true parameters α and β which determine the linear relationship between the unknown expected measurement values. The residual variances $\sigma_{r_i}^2$ are therefore approximately equivalent to the measurement error variances and thus not necessarily constant over the measuring range. Note again, that is is assumed here, that the residuals are statistically independent, which is not an appropriate assumption for a parametric regression model (PCA or SPCA).

6.2 The Concept of Local Confidence Intervals

As pointed out in Section 5.2, outlier limits can be given in form of robust confidence intervals for the orthogonal residuals. Since the orthogonal residual variance is not necessarily constant, a local residual variance estimator is needed. The local residual variance will be estimated from all observed residuals $r_1, ..., r_n$. Each residual will be weighted concerning the information it contains for the specific local residual variance estimate under consideration. The variance estimator will hence be constructed as the sum of weighted residuals with appropriate weights w_{ik}. The LORELIA Residual Variance Estimator will be given as:

$$\hat{\sigma}_{r_i}^2 = \frac{1}{\sum_{l=1}^n w_{il}} \cdot \sum_{k=1}^n w_{ik} \cdot r_k^2, \quad \text{for } i = 1, ..., n. \tag{6.2.1}$$

Note that (6.2.1) is based on the assumption that:

$$E(R_i) = 0, \text{ for all } i = 1, ..., n.$$

[Satterthwaite, 1941] pointed out that a complex variance estimator like (6.2.1), will have approximately the following distributional property:

$$\frac{DF_i \cdot \hat{\sigma}_{E_i}^2}{\sigma_{E_i}^2} \overset{\text{approx}}{\sim} \chi_{DF_i}^2, \text{ for } i = 1, ..., n,$$

where DF_i are the corresponding degrees of freedom calculated from the formula:

$$DF_i = \frac{\left(\frac{1}{\sum_{k=1}^n w_{ik}} \cdot \sum_{k=1}^n w_{ik} \cdot r_k^2\right)^2}{\frac{1}{\left(\sum_{k=1}^n w_{ik}\right)^2} \cdot \sum_{k=1}^n w_{ik}^2 \cdot r_k^4} = \frac{\left(\sum_{k=1}^n w_{ik} \cdot r_k^2\right)^2}{\sum_{k=1}^n w_{ik}^2 \cdot r_k^4}, \quad \text{for } i = 1, ..., n, \tag{6.2.2}$$

which is given in ([Satterthwaite, 1941], page 313) and ([Satterthwaite, 1946], page 111). Note that in [Qian, 1998] it is shown that this formula may underestimate the effective degrees of freedom. Some approaches to correct this downward bias are discussed, which are however not applied in the context of this work for the sake of simplicity.

By (6.1.5), it holds that:

$$\frac{R_i}{\sigma_{r_i}} \overset{iid}{\sim} N(0, 1), \text{ for } i = 1, ..., n.$$

Thus, by (6.2.2) and (6.2.3) it follows with [Fahrmeir et al., 2007] (Chapter B, Page 461) that:

$$\sqrt{DF_i} \cdot \frac{R_i}{\sigma_{r_i}} \cdot \frac{\sigma_{r_i}}{\sqrt{DF_i} \cdot \hat{\sigma}_{r_i}} = \frac{R_i}{\hat{\sigma}_{r_i}} \overset{\text{approx}}{\sim} t_{DF_i}, \text{ for } i = 1, ..., n. \tag{6.2.3}$$

With (6.2.3) it is easy to deduce a $(1-\alpha)\%$ approximative confidence interval for the ith orthogonal residual:

$$C_{\alpha,i} := [-t_{DF_i,(1-\frac{\alpha}{2})} \cdot \hat{\sigma}_{r_i}, t_{DF_i,(1-\frac{\alpha}{2})} \cdot \hat{\sigma}_{r_i}], \quad \text{for } i = 1, ..., n. \tag{6.2.4}$$

The confidence intervals (6.2.4) will be used as the local outlier limits in the following. The definition of these outlier limits implies that outliers correspond to residuals which are very unlikely to come from the corresponding normal distribution $N(0, \sigma_{r_i}^2)$.

Each residual r_i which lays outside its corresponding confidence interval $C_{\alpha,i}$ will be identified as an outlier. This correspond to a multiple test situation as described in Section 3.1.2. In order to keep a global significance level of α_{glob}, the local significance levels has to be adjusted. Thus every residual r_i is compared to $C_{\alpha_{\text{loc}_i},i}$ where the α_{loc_i}'s are determined by the chosen adjustment procedure.

6.3 How to Weight - Newly Developed Criteria

The new outlier test is based on the construction of local confidence limits, which depend on a weighted residual variance estimator. Hence, the main task of this work will be to construct powerful weights. The new weighting method of the LORELIA Residual Test go back to a weighting procedure proposed in [Konnert, 2005] which will be presented in Section 6.3.1. In Section 6.3.2, new concept for the construction of improved weights will be presented.

6.3.1 Historical Background - Basic Ideas

The weights proposed in [Konnert, 2005] are given by:

$$w_{ik}^{\text{Kon}} := \frac{1}{1 + \delta_{ik}}, \quad \text{for } i, k = 1, ..., n, \tag{6.3.1}$$

where

$$\delta_{ik} := (x_i^p - x_k^p)^2 + (y_i^p - y_k^p)^2, \quad \text{for } i, k = 1, ..., n$$

and (x_i^p, y_i^p) is the orthogonal projection of (x_i, y_i) to the regression line, given by:

$$x_i^p = \frac{x_i + \widehat{\beta}_{PB}(y_i - \widehat{\alpha}_{PB})}{1 + \widehat{\beta}_{PB}^2}, \quad y_i^p = \widehat{\beta}_{PB} \cdot x_i^p + \widehat{\alpha}_{PB}, \quad \text{for } i = 1, ..., n.$$

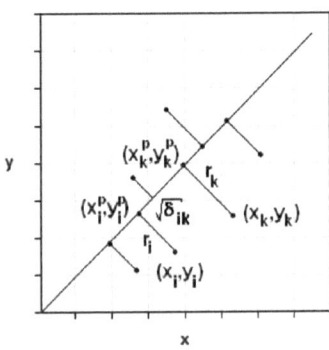

Figure 6.2: Distance between the Orthogonal Residuals

By (6.3.2), the δ_{ik}'s are given as the squared distances between the orthogonal projections to the regression line. The weights w_{ik}^{Kon} generally decrease with increasing distances δ_{ik}. It is easy to see that $w_{ik}^{\text{Kon}} \in (0,1]$ for all $i,k = 1,...,n$ and that:

$$w_{ik}^{\text{Kon}} \to 1 \quad \text{as} \quad \delta_{ik} \to 0,$$
$$w_{ik}^{\text{Kon}} \to 0 \quad \text{as} \quad \delta_{ik} \to \infty, \quad \text{for all } i,k = 1,...,n.$$

The definition of the weights in (6.3.1) is strongly related to the 'Inverse Distance Method', also called the 'Shepard's Method' proposed by [Shepard, 1968], which is used for interpolation. The Shepard's weights are given as:

$$w_{ik}^{\text{Shep}} := \frac{1}{\delta_{ik}}, \quad \text{for} \quad i,k = 1,...,n.$$

Note that, for the interpolation problem it always holds that $\delta_{ik} \neq 0$ for $i,k = 1,...,n$, unlike for the weighting problem in the context of this work.

The residual variance estimator based on the weights defined in (6.3.1) can be biased by the presence of outliers. In order to protect against masking effects, [Konnert, 2005] proposed to neglect the largest term of $w_{ik}^{\text{Kon}} \cdot r_k^2$ in formula (6.2.1). With:

$$w_{m_i} := \max\left\{w_{i1} \cdot r_1^2, w_{i2} \cdot r_2^2, ..., w_{in} \cdot r_n^2\right\}. \tag{6.3.2}$$

the residual variance estimator thus becomes:

$$\hat{\sigma}_{r_i}^2 = \frac{1}{\sum_{\substack{k=1 \\ k \neq m_i}}^{n} w_{ik}^{\text{Kon}}} \cdot \sum_{\substack{k=1 \\ k \neq m_i}}^{n} w_{ik}^{\text{Kon}} \cdot r_k^2, \text{ for } i = 1,...,n. \tag{6.3.3}$$

6.3.1.1 Problems and Limitations

The weights proposed in (6.3.1) are based on the distance measure δ_{ik}. This insures that that the residual variance is *locally* estimated as the weights decrease with increasing distance between the corresponding orthogonal projections. However there exists severe disadvantages and problems coming along with this definition of the weights, which will be explored in the following:

(i.) A major disadvantage of the above weighting method is the fact that it is not robust against outliers. Although the term $w_{im_i} \cdot r_{m_i}^2$ defined in (6.3.2) is excluded in the calculation of (6.2.1) to avoid masking effects, this does not necessarily contribute to more stable residual variance estimates: On the one hand, (x_{m_i}, y_{m_i}) does not necessarily have to be an outlier, on the other hand, there may be more then one outlier in the neighborhood of the ith residual. To visualize the above problems, consider the Passing-Bablok regression fit and the corresponding residual plot for the following simulated dataset. The local outlier limits determined by the weights defined by [Konnert, 2005] are marked by the gray lines.

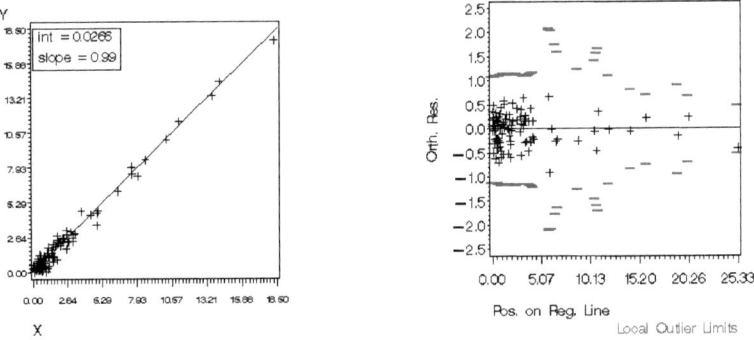

Figure 6.3: The Method of A. Konnert for a Dataset With No Obvious Outlier

No outliers are identified as no measurement value lays outside its corresponding confidence interval. Note, that the confidence limits are not constant over the measuring range although a constant residual variance has been simulated. The local data density seems to have a high influence on the actual residual variance estimates. The local confidence intervals do not merge smoothly - no global trend of the outlier limits can be deduced.

Now, consider the same dataset with one simulated outlier at the lower concentration range which is marked in light gray.

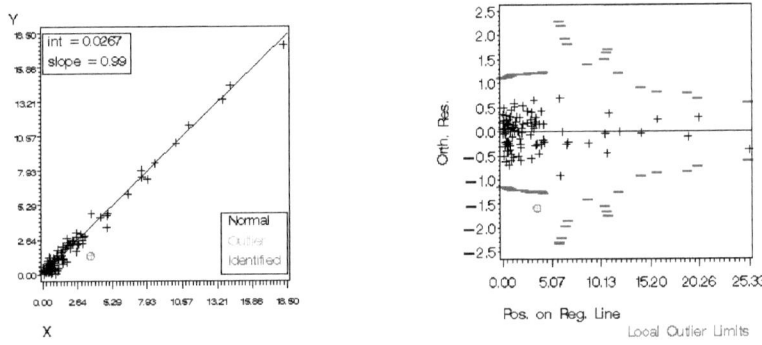

Figure 6.4: The Method of A. Konnert for the Dataset with One Outlier

The circle indicates that the simulated outlier is well identified by the test. The confidence limits look very similar to those for the data set without the outlier. Since the outlier occurs in a high information density area, the negligence of $w_{im_i} \cdot r_{m_i}^2$ avoids that the presence of the outlier bias the surrounding residual variance estimates. However, if a second outlier at the lower concentration range is added, the variance estimates get strongly biased:

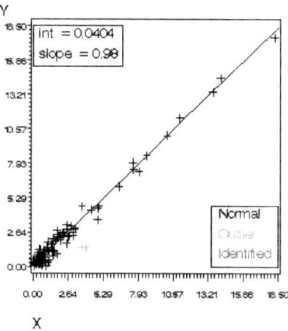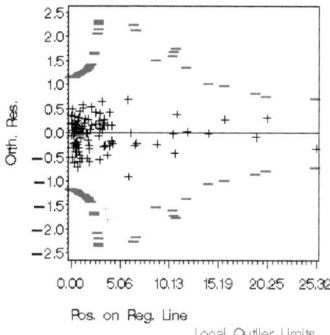

Figure 6.5: The Method of A. Konnert for the Dataset with Two Neighbored Outliers

Now, both outliers are not identified anymore. The confidence limits at the lower concentration range are much wider now than in the dataset with one or no outlier. At the higher concentration range however, the width of the local confidence intervals has hardly changed.

(ii.) Another deficiency is that the weights are not invariant under axes scaling, unlike the similar Shepard's method. A change of the unit of measurement in both methods will lead to different values for the weights and thus to different variance estimates.

Theorem 6.1
Let I_1 be the measuring range of a given dataset. A scaled measuring range of I_1 is defined by $I_2 := F \cdot I_1$ with $F > 1$. Let w_{ik,I_1}^{Kon} and w_{ik,I_2}^{Kon} denote the weights defined in (6.3.1) with respect to I_1 and I_2, respectively. Then, the weights for the original measuring range I_1 and for the scaled measuring range I_2 are related by:

$$w_{ik,I_1}^{\text{Kon}} = w_{ik,I_2}^{\text{Kon}} = 1, \text{ if } i = k, \tag{6.3.4}$$

$$w_{ik,I_1}^{\text{Kon}} < w_{ik,I_2}^{\text{Kon}}, \quad \text{if } i \neq k, \quad \text{for } i = 1, ..., n. \tag{6.3.5}$$

Proof: Equation (6.3.4) follows directly by the definition of the weights given in [Konnert, 2005]. Now, choose $i \neq k$ for $i, k \in \{1, ..., n\}$. For a scaling factor $F > 1$ it holds:

$$w_{ik,I_1}^{\text{Kon}} = \frac{1}{1 + \delta_{ik,I_1}} < \frac{1}{1 + F \cdot \delta_{ik,I_1}} < \frac{1}{1 + \delta_{ik,F \cdot I_1}} = \frac{1}{1 + \delta_{ik,I_2}} = w_{ik,I_2}^{\text{Kon}}.$$

□

By Theorem 6.1, the residual variance estimator proposed by [Konnert, 2005] is not invariant under axes scaling. Moreover with (6.3.4) and (6.3.5) it can easily be deduced that the relative influence of the ith residual to the ith residual variance estimator given by $\frac{w_{ii,F \cdot I_1}}{\sum_{l=1}^{n} w_{il,F \cdot I_1}}$ is an increasing function of F. For a large scaling factor F, the local residual variance estimate $\hat{\sigma}_{r_i}^2$ will thus mainly be influenced by the ith residual itself. As a consequence, a measuring range with large units in absolute numbers will lead to more local variance estimates. The following graph shows the local outlier limits for the weighting method of [Konnert, 2005] for an exemplary dataset with the original measuring range, a 10times and a 100times scaled measuring range:

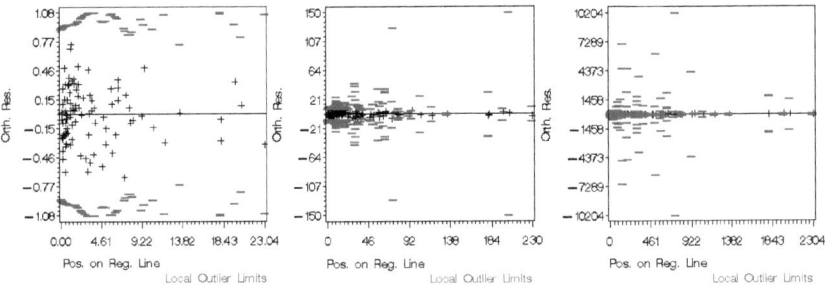

Figure 6.6: Local Outlier Limits for Scaled Measuring Ranges

The larger the scale factor of the measuring range, the less smooth are the joints between neighbored local outlier limits.

(iii.) Now, consider the properties of the above weighting method for a dataset with an inhomogeneous data distribution, as it often occurs in practical application (compare for example the dataset given in Figure 6.3). In an area with low data density, the ith residual variance estimate will be mainly influenced by the ith residual itself. Hence the value of the residual variance estimator will be close to r_i^2. Therefore, the ith residual is very likely to fall within the corresponding confidence interval given by (6.2.4).

However, if the data density is low, it is also likely to happen that w_{im_i} defined in (6.3.2) is given by $w_{im_i} = w_i$, even if r_i is not an outlier! If r_i is not an outlier and does not contribute to the ith orthogonal residual variance estimate, in an area with low data density this may lead to a seriously biased estimate of $\sigma_{r_i}^2$ and hence to a wrong outlier classification.

Generally, for areas with low data density, the local variance estimates will not be stable since there is not much local data evidence. In dense data regions, closely neighbored residuals have a high influence on the variance estimate which will lead to more stable estimates. In this case however, the variance estimates can easily get biased by the presence of several outliers, as it has been shown in the above examples.

6.3.2 New Concepts for Weight Construction

In this section new concept for the construction of powerful weights will be presented. The basic ideas by [Konnert, 2005] are improved and new ideas are developed.

6.3.2.1 Construction of a Local Estimator

In order to construct a *local* residual variance estimator, the residual variance estimate $\hat{\sigma}_{r_i}^2$ should be influenced more by the residuals which are closely neighbored to r_i than by those located far away. As a consequence, a measure for the distance between the residuals is needed. Since the residuals are defined as the orthogonal distances to the regression line, a distance measure for the orthogonal projections will be considered.

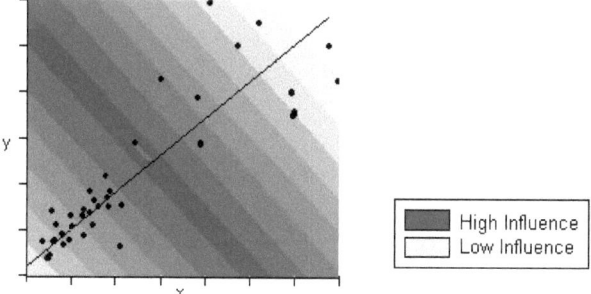

Figure 6.7: Influence of the Neighbored Residuals

The weights proposed in [Konnert, 2005] are exclusively based on a simple distance measure. This insures that the residual variance estimator becomes local. However, there exist several very important other requirements for an efficient weighting function, which will be presented in the following.

6.3.2.2 Construction of an Outlier Robust Estimator

In order to achieve a residual variance estimator, which is robust against outliers, each residual should be weighted according to its local reliability. The local reliability of a given residual r_k is high if the closely neighbored residuals are of the same magnitude. Residuals located far away from r_k may generally be of a very different magnitude and should therefore be considered less important when the reliability of r_k is judged. The task is to construct a reliability measure which is low for locally surprisingly large residuals. In this case, all neighbored residuals are of much smaller magnitude. A reliability measure may therefore be based on the sum of distance-weighted residual differences as it will be explored in Section 6.4.2.

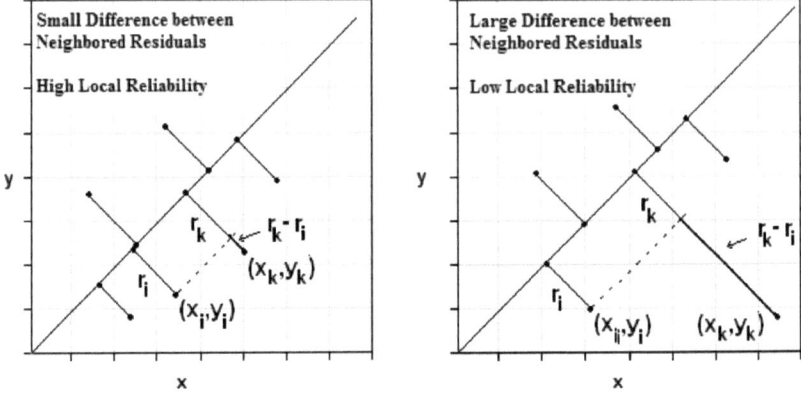

Figure 6.8: The Local Reliability

In the above figure, the concepts of a high and a low local reliability are visualized.

6.3.2.3 Invariance under Axes Scaling

A change in the unit of measurement for both methods should have no influence on the values of the outlier limits. In order to achieve this, the weights should be invariant under axes scaling. The most simple idea would be to standardize the measuring range in advance to the calculation of the local outlier limits. Another idea is to construct *relative* measures for the distance and for local reliability rather than absolute measures. In this work, the second approach is chosen.

The constructed relative measures defined in Sections 6.4.1 and 6.4.2 involve the additional information of the sample size which is neglected in the approach of [Konnert, 2005]. The influence of the sample size is especially important to fulfill the requirements described in Section 6.3.2.5.

6.3.2.4 The Meaning of the Local Data Information Density

The data density can differ a lot between different datasets and may even be very inhomogeneous within a single dataset. Thus, the (local) information density differs between and within datasets. A lower information density should always correspond to wider outlier limits and thus to a more conservative outlier test, as the level evidence for the outlier classification is low. On the other hand, in densely distributed data regions, where the information density and hence the level of evidence is high, existing outliers should be more easily identified.

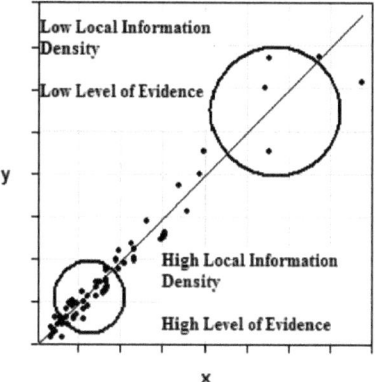

Figure 6.9: Different Areas of Information Density

6.3.2.5 The Co-Domain of the Weights

The global co-domain W_n of the weights is determined by the highest and the lowest possible weight for an arbitrary dataset with sample size n:

$$W_n := \left[\inf_{\{i,k=1,...,n\}} \{w_{ik}\} , \sup_{\{i,k=1,...,n\}} \{w_{ik}\} \right] , \text{ for an } i = 1,...,n.$$

The range of observed weights for a specific dataset $\left[\min_{\{i,k=1,...,n\}} \{w_{ik}\} , \max_{\{i,k=1,...,n\}} \{w_{ik}\} \right]$ is always a subset of W_n. The width of W_n should depend on the sample size n. For small sample sizes, the amount of available information is limited. Therefore, none of the observed residual should be radically down weighted. For high sample sizes, there is enough information available to neglect a few of the observed residuals. Therefore, the width of the global co-domain W_n should increase with increasing sample size n.

6.4 The Weights for the LORELIA Residual Test

In this section the weighting method for the new LORELIA Residual Test is presented. The weights for the LORELIA Residual Test are given as the product of a distance measure Δ_{ik} and a measure for the local reliability $\Gamma_{k,n}$ which depends on the sample size n as required in Sections 6.3.2.1 and

6.3.2.2:

$$w_{ik} := \Delta_{ik} \cdot \Gamma_{k,n}, \quad \text{for } i,k = 1,...,n. \tag{6.4.1}$$

The distance measure Δ_{ik} ensures, that the residual variance is *locally* estimated. The reliability measure $\Gamma_{k,n}$ is needed in order to construct a residual variance estimator which is robust against outliers.

The distance and the reliability measure will be explicitly defined in Sections 6.4.1 and 6.4.2. The properties of the new weights will be mathematically explored and it will be shown that all the requirements given in Section 6.3.2 are fulfilled.

Note however, that there exist multiple possibilities to construct weights based on the principle ideas presented in Section 6.3.2. Therefore, the weighting method for the LORELIA Residual Test which has been developed in the context of this work should not be considered as an exclusive solution to the general problem but may be further developed and improved.

6.4.1 Definition of the Distance Measure

In [Konnert, 2005] the squared distances δ_{ik}^2 between the orthogonal projections are proposed as a distance measure. The δ_{ik}'s are transformed in the following way:

$$\frac{1}{\delta_{ik}+1}, \quad \text{for } i,k = 1,...,n \tag{6.4.2}$$

in order to achieve a co-domain of $[0,1]$ for the weights (compare Section 6.3.1). As it has been already explored, this distance measure does not meet the requirement to be invariant under axes scaling as claimed in Section 6.3.2.3. To achieve this, the absolute distances δ_{ik} are replaced by relative distances with respect to the mean distance. The mean distance is given by:

$$\bar{\delta} = \frac{1}{n^2} \sum_{i=1}^{n} \sum_{k=1}^{n} \delta_{ik}. \tag{6.4.3}$$

The considered relative distances are thus given as:

$$\frac{\delta_{ik}}{\bar{\delta}}, \quad \text{for } i,k = 1,...,n. \tag{6.4.4}$$

The new LORELIA Distance Measure fulfills the requirements given in Section 6.3.2 as shown in the following:

Definition 6.2 (The LORELIA Distance Weight)

The LORELIA Distance Weight is defined as:

$$\Delta_{ik} := \frac{1}{\frac{\delta_{ik}}{\delta} + 1}, \quad \text{for } i, k = 1, ..., n. \tag{6.4.5}$$

It is easy to see that:

$$\Delta_{ii} = 1, \text{ for } i = 1, ..., n.$$

and

$$0 < \Delta_{ik} < 1, \text{ for } i \neq k, \ i, k = 1, ..., n.$$

The new distance measure (6.4.5) has the following important properties:

Theorem 6.3

The LORELIA Distance Measure Δ_{ik} defined in Definition 6.2 is invariant under axes scaling.

Proof: If I_1 is the measuring range of a given dataset and $I_2 := F \cdot I_1, F > 1$ is a scaled measuring range, it holds:

$$\begin{aligned}
\Delta_{ik,I_2} &= \frac{1}{\frac{n^2 \delta_{ik,I_2}}{\sum_{l=1}^{n} \sum_{m=1}^{n} \delta_{lm,I_2}} + 1} \\
&= \frac{1}{\frac{n^2 \cdot \delta_{ik,F \cdot I_1}}{\sum_{l=1}^{n} \sum_{m=1}^{n} \delta_{lm,F \cdot I_1}} + 1} \\
&= \frac{1}{\frac{n^2 \cdot F \cdot \delta_{ik,I_1}}{\sum_{l=1}^{n} \sum_{m=1}^{n} F \cdot \delta_{lm,I_1}} + 1} \\
&= \frac{1}{\frac{n^2 \delta_{ik,I_1}}{\sum_{l=1}^{n} \sum_{m=1}^{n} \delta_{lm,I_1}+1} + 1} \\
&= \Delta_{ik,I_1}, \qquad\qquad \text{for } i, k = 1, .., n.
\end{aligned}$$

$\square$

The new distance measure Δ_{ik} takes account of different data densities as required in Section 6.3.2.4. In areas with low data density the distances between the residuals are large, thus all distance weights Δ_{ik} with $k \neq i$ are much smaller than $\Delta_{ii} = 1$. In high information density areas however, all residuals r_k which are closely neighbored to r_i will have distance weights $\Delta_{ik} \approx 1$.

The following theorem insures that the requirements of Section 6.3.2.5 are met:

Theorem 6.4

The co-domain for the LORELIA Distance Weights is given by:

$$\left[\inf_{\{i,k=1,\ldots,n\}}\{\Delta_{ik}\}\,,\ \sup_{\{i,k=1,\ldots,n\}}\{\Delta_{ik}\}\right] = \left[\frac{1}{\frac{n^2}{2\cdot(n-1)}+1},\,1\right]. \quad (6.4.6)$$

Proof: It is easy to see that:

$$\sup_{\{i,k=1,\ldots,n\}}\{\Delta_{ik}\} = 1.$$

In order to determine:

$$\inf_{\{i,k=1,\ldots,n\}}\{\Delta_{ik}\}$$

calculate:

$$\lim_{\delta_{ik}\to\infty}\Delta_{ik},\quad \text{for given } i,k=1,\ldots,n.$$

Without loss of generality, one may as well calculate:

$$\lim_{\delta_{(i-1)i}\to\infty}\Delta_{(i-1)i},\quad \text{for a given } i=2,\ldots,n,$$

as illustrated in Figure 6.10:

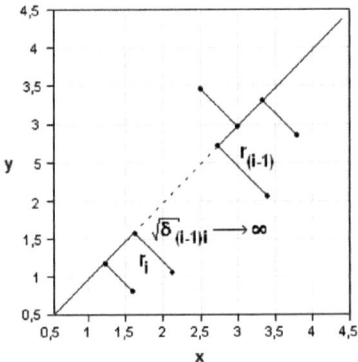

Figure 6.10: Increasing Distance $\delta_{(i-1)i}$

It holds:

$$\frac{n^2 \cdot \delta_{(i-1)i}}{\sum_{s=1}^{n} \sum_{t=1}^{n} \delta_{st}}$$

$$= \frac{n^2 \cdot \delta_{(i-1)i}}{2\sum_{s=1}^{i-1} \sum_{t=i}^{n} \delta_{st} + \sum_{s=1}^{i-1} \sum_{t=1}^{i-1} \delta_{st} + \sum_{s=i}^{n} \sum_{t=i}^{n} \delta_{st}}$$

$$= \frac{n^2 \cdot \delta_{(i-1)i}}{2\left(\sum_{s=1}^{i-1} \sum_{t=i}^{n} \delta_{(i-1)i} + \sum_{s=1}^{i-1} \sum_{t=i}^{n} \left(\delta_{(i-1)s} + \delta_{it}\right)\right) + \sum_{s=1}^{i-1} \sum_{t=1}^{i-1} \delta_{st} + \sum_{s=i}^{n} \sum_{t=i}^{n} \delta_{st}}$$

$$= \frac{n^2}{2\sum_{s=1}^{i-1} \sum_{t=i}^{n} 1 + 2\sum_{s=1}^{i-1} \sum_{t=i}^{n} \underbrace{\frac{\delta_{(i-1)s} + \delta_{it}}{\delta_{(i-1)i}}}_{\to\, 0 \text{ for } \delta_{(i-1)i} \to \infty} + \underbrace{\frac{\left(\sum_{s=1}^{i-1} \sum_{t=1}^{i-1} \delta_{st} + \sum_{s=i}^{n} \sum_{t=i}^{n} \delta_{st}\right)}{\delta_{(i-1)i}}}_{\to\, 0 \text{ for } \delta_{(i-1)i} \to \infty}}$$

$$\to \frac{n^2}{2(n-i+1)(i-1)}, \quad \text{as } \delta_{(i-1)i} \to \infty, \text{ for } i = 2, ..., n.$$

Therefore:

$$\Delta_{(i-1)i} \xrightarrow{\delta_{(i-1)i} \to \infty} \frac{1}{\frac{n^2}{2\cdot(n-i+1)(i-1)} + 1}$$

$$\geq \frac{1}{\frac{n^2}{2\cdot\min_{\{i=2,...,n\}}\{(n-i+1)\cdot(i-1)\}} + 1}$$

$$= \frac{1}{\frac{n^2}{2(n-1)} + 1}. \tag{6.4.7}$$

Thus, the co-domain for the Δ_{ik}'s is given as:

$$\left[\inf_{\{i,k=1,...,n\}}\{\Delta_{ik}\}, \sup_{\{i,k=1,...,n\}}\{\Delta_{ik}\}\right] = \left[\frac{1}{\frac{n^2}{2\cdot(n-1)} + 1}, 1\right].$$

$\square$

Therefore, it holds:

$$\left[\inf_{\{i,k=1,...,n\}}\{\Delta_{ik}\}, \sup_{\{i,k=1,...,n\}}\{\Delta_{ik}\}\right] \to [0,1], \quad \text{for } n \to \infty. \tag{6.4.8}$$

With Theorem 6.3, it can easily be seen that the co-domain of the distance weights Δ_{ik} and thus also the co-domain of the overall weights $w_{ik} = \Delta_{ik} \cdot \Gamma_{k,n}$ is dependent of the sample size n. Thus, the requirement given in Section 6.3.2.5 is fulfilled.

To illustrate the concept, consider for example the smallest possible distance weight for a sample size of $n = 10$:

$$\frac{1}{\frac{10^2}{2\cdot(10-1)} + 1} \approx 0.1525,$$

which is about 10 times higher than the smallest possible weight for a sample size of $n = 100$:

$$\frac{1}{\frac{100^2}{2 \cdot (100-1)} + 1} \approx 0.0194.$$

Hence, for a sample size of $n = 100$, the weights can scatter in a wider range than for a sample size of $n = 10$. In Figure 6.11, the squared absolute distances δ_{1k}^2 to the 1st residual are plotted against Δ_{1k} for a dataset with $n = 100$ observations and a reduced dataset with $n = 10$ observations. Since the second dataset is only a reduced version of the first, the absolute distances between data values remain the same if the fact that the regression parameters will change slightly is neglected.

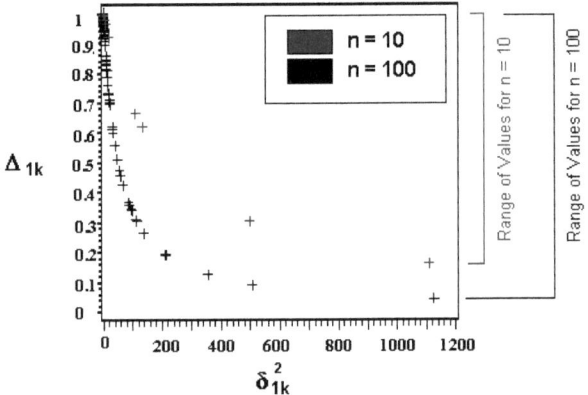

Figure 6.11: The Values of the Distance Measure Δ_{ik} for Different Sample Sizes

Although the absolute distances δ_{1k} between the residuals remain the same for corresponding measurement values, the actual range of values for the distance weights Δ_{1k} is narrowed by the reduction of the sample size.

6.4.2 Definition of a Reliability Measure

In this section a measure for the local reliability, as required in Section 6.3.2.2, will be constructed. The local reliability of a fixed residual r_k is high, if closely neighbored residuals are of the same magnitude. The local reliability of r_k is low, if the surrounding residuals are much smaller than r_k (compare Figure 6.8). A reliability measure may therefore be based on the weighted sum of differences between each residual and r_k, where the weights are given by the LORELIA Distance Measure Δ_{ik}:

$$\sum_{l=1}^{n} \Delta_{lk} \cdot (|r_l| - |r_k|)^2, \; k = 1, ..., n.$$

In order to construct a local reliability measure, which is invariant under axes scaling, as claimed in Section 6.3.2.3, consider the relative sum of weighted differences between the residuals:

$$\gamma_{k,n} := \frac{\sum_{l=1}^{n} \Delta_{lk} \cdot (|r_l| - |r_k|)^2}{\sum_{m=1}^{n} \sum_{s=1}^{n} \Delta_{ms} \cdot (|r_m| - |r_s|)^2}, \text{ for } k = 1, ..., n. \qquad (6.4.9)$$

Note that the reliability measure $\gamma_{k,n}$ depends on the sample size n. The values of $\gamma_{k,n}$ generally decrease as n increases.

If r_k is an outlier, $\gamma_{k,n}$ will be close to 1. For a residual which is of the same magnitude as the surrounding residuals, $\gamma_{k,n}$ will be close to 0. The influence of the distance measure Δ_{lk} in (6.4.9) guarantees, that the reliability is *locally* measured. Note that the local reliability $\gamma_{k,n}$ of a residual r_k is the same for every residual variance estimate $\hat{\sigma}_i^2$ under consideration and thus does not depend on i. The LORELIA Reliability Weight $\Gamma_{k,n}$ will be given as a function of $\gamma_{k,n}$:

Definition 6.5 (The LORELIA Reliability Weight)
For a constant parameter $c > 1$ the LORELIA Reliability Weight is defined as:

$$\Gamma_{k,n} := \begin{cases} 1, & \text{for } \gamma_{k,n} \leq \frac{1}{n} \\ 0.5 \cdot \left[\cos\left((\gamma_{k,n} - \frac{1}{n}) \cdot \frac{\pi}{\frac{c}{n+(c-1)} - \frac{1}{n}} \right) + 1 \right] & \text{for } \frac{1}{n} < \gamma_{k,n} < \frac{c}{n+(c-1)} \\ 0, & \text{for } \gamma_{k,n} \geq \frac{c}{n+(c-1)} \end{cases} \qquad (6.4.10)$$

Thereby, c is a parameter to adjust the robustness of the outlier test for the given data situation.

The LORELIA Reliability Weight $\Gamma_{k,n}$ is based on the following considerations:

(i.) If the residual variances are assumed to be constant over the measuring range and no outliers are present, then $\gamma_{k,n} \approx \frac{1}{n}$ for all $k = 1, ..., n$. Hence, all residuals with $\gamma_{k,n} \leq \frac{1}{n}$ will get a reliability weight of 1.

(ii.) If one residual out of n is c times larger or more than the remaining $n - 1$, this residual will get the minimal reliability weight of 0. Thus, this residual does not influence the variance estimates at all. The choice of the worst case limit $\frac{c}{n+(c-1)}$ is important to adjust the robustness of the variance estimator. If c is chosen too large, this will result in false negative test results. Existing outliers may not be detected as only *very* extreme residuals are classified as outliers. However, if c is chosen too small, the reliability measure becomes too sensible for the normal data scattering which will result in many false positive outlier identifications. Several values for the limit c has been tested by the author in various data situations. Due to these experimental studies, it is recommended to use a limiting value of:

$$c = 10. \qquad (6.4.11)$$

However, this value may not be appropriate for unusual or extreme data situations. It will be the task of the data analyst to adjust the value of c in this case. In the context of this work, the LORELIA Residual Test is always applied with $c = 10$.

(iii.) For residuals r_k with $\frac{1}{n} < \gamma_{k,n} < \frac{c}{n+(c-1)}$, the function

$$0.5 \cdot \left[\cos\left((\gamma_{k,n} - \frac{1}{n}) \cdot \frac{\pi}{\frac{c}{n+(c-1)} - \frac{1}{n}} \right) + 1 \right]$$

is chosen to insure that $\Gamma_{k,n}$ is a continuously differentiable, decreasing function in $\gamma_{k,n}$ over $[0,1]$, which is point symmetric in

$$\left(0.5 \cdot \left(\frac{1}{n} + \frac{c}{n+(c-1)} \right), f\left(0.5 \cdot \left(\frac{1}{n} + \frac{c}{n+(c-1)} \right) \right) \right).$$

In the following graph, $\Gamma_{k,n}$ is plotted as a function of $\gamma_{k,n}$ for different sample sizes:

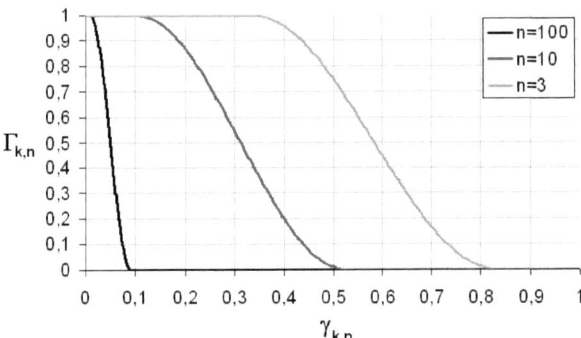

Figure 6.12: The Local Reliability Measure $\Gamma_{k,n}$ for Different Sample Sizes and $c = 10$

6.5 Definition of the LORELIA Residual Test

In this section, the LORELIA Residual Test is summarized and formally defined to give a general overview of the new test. The motivation for the given formulas and definitions are given in the previous sections and will not be repeated here.

The new LORELIA Residual Test is explicitly formulated as follows:

Definition 6.6 (The LORELIA Test Hypotheses)
The LORELIA Residual Test is based on the following test hypotheses:

$$H_0 : \text{The considered dataset contains no outliers,}$$
$$\text{versus} \quad (6.5.1)$$
$$H_1 : \text{The considered dataset contains outliers.}$$

H_0 is rejected if any of the observed residuals r_i exceeds the given outlier limits:

$$\text{Reject } H_0 \Leftrightarrow \exists i \in \{1, ..., n\} : r_i \notin C_{\alpha_{\text{loc}},i}, \quad (6.5.2)$$

where $C_{\alpha_{\text{loc}},i}$ is an $(1 - \alpha_{\text{loc}})\%$ approximate local confidence interval for the residual R_i.

As the multiple test situation (6.5.2) will lead to the accumulation of first order errors, the local level of significance α_{loc} has to be adjusted by an appropriate adjustment method in order to keep a global significance level of α_{glob}. A general discussion on this task has been given in Section 3.1.2. Considerations on the choice of the adjustment methods for the LORELIA Residual Test will be given in Section 7.4. An overview of different adjustment methods can be found in [Hochberg, Tamhane, 1987] and [Hsu, 1996].

Outlier limits for the LORELIA Residual Test are given as local confidence intervals $C_{\alpha_{\text{loc}},i}$ for the orthogonal residuals. These confidence intervals are defined as follows:

Definition 6.7 (The LORELIA Outlier Limits)
The LORELIA Outlier Limits are given by:

$$C_{\alpha,i} := [-t_{DF_i,(1-\frac{\alpha_{\text{loc}}}{2})} \cdot \hat{\sigma}_i, t_{DF_i,(1-\frac{\alpha_{\text{loc}}}{2})} \cdot \hat{\sigma}_i], \quad \text{for } i = 1, ..., n. \quad (6.5.3)$$

where $\hat{\sigma}_{r_i}^2$ is a local residual variance estimator defined as:

$$\hat{\sigma}_{r_i}^2 = \frac{1}{\sum_{l=1}^{n} w_{il}} \cdot \sum_{k=1}^{n} w_{ik} \cdot r_k^2, \quad \text{for } i = 1, ..., n. \quad (6.5.4)$$

and $t_{DF_i,(1-\frac{\alpha_{\text{loc}}}{2})}$ is the $(1 - \alpha_{\text{loc}})\%$ quantile of the Student's-t distribution with DF_i degrees of freedom calculated from the formula:

$$DF_i = \frac{(\sum_{k=1}^{n} w_{ik} \cdot r_k^2)^2}{\sum_{k=1}^{n} w_{ik}^2 \cdot r_k^4}, \quad \text{for } i = 1, ..., n. \quad (6.5.5)$$

The LORELIA Residual Variance Estimator is based on the following weighting function:

Definition 6.8 (The LORELIA Weights)
The LORELIA Weights are defined as:

$$w_{ik} := \Delta_{ik} \cdot \Gamma_{k,n}, \quad \text{for } i, k = 1, ..., n, \qquad (6.5.6)$$

where Δ_{ik} is a measure for the distance between r_i and r_k along the regression line ,defined in Definition 6.2 in Section 6.4.1, and $\Gamma_{k,n}$ is a measure for the local reliability of r_k which depends on the sample size n and is defined in Definition 6.5 in Section 6.4.2.

With the above definitions the LORELIA Residual Test is entirely defined.

Chapter 7

Performance of the New LORELIA Residual Test

In this chapter, the performance of the LORELIA Residual Test will be evaluated in order to answer the following questions:

- In real data situations, does the LORELIA Residual Test identifies visually suspicious values as outliers?

- For simulated data situations, are simulated outliers truly identified? How many outlier misclassifications are there?

- In the case of Bonferroni adjusted outlier limits, does the new test meet its predefined significance level?

- What are the performance differences between the new LORELIA Residual Test and the common outlier tests presented in Chapter 5?

- How good is the new test in standard data situations which can be handled with other outlier tests?

- How good is the LORELIA Residual Test for more complex data situations in which standard outlier tests will fail?

- Which factors influence the performance of the new test?

- Which problems and limitations can occur and how can they be handled?

- How does the choice of the adjustment procedure influence the performance of the LORELIA Residual Test?

The new LORELIA Residual Test will be used on a variety of different datasets in order to answer the above questions. Thereby, the performance of the test strongly depends on the underlying data situation. Different datasets are classified concerning the following criteria:

(i.) The distribution of the measurement values within the measuring range,

(ii.) The underlying residual variance model,

(iii.) The number of outliers in the dataset,

(iv.) The magnitude of the outlier terms,

(v.) The position and the distribution of outliers.

Throughout this chapter, the LORELIA Residual Test is used for a global significance level of $\alpha_{\text{glob}} = 0.1$ which is adjusted with the conservative Bonferroni procedure to insure that the type 1 error is at most α_{glob} as measurement values which are wrongly identified as outlier cause much additional work and troubles. The local significance level is thus given by $\alpha_{\text{loc}} = \frac{\alpha_{\text{glob}}}{n}$.

To begin with, in Section 7.1, the LORELIA Residual Test will be compared to the common outlier tests for method comparison studies proposed by [Wadsworth, 1990], which are presented in Chapter 5.

In Section 7.1.1, all tests will be applied on a variety of exemplary datasets in order to give a first impression on the performance of the different tests. These datasets represent common data situations which can be met in clinical practice. It will be checked visually which test identifies suspicious values best and if the calculated outlier limits seem appropriate.

As the test performance strongly depends on the underlying data situations, a general ranking of the different outlier tests is not possible. However, the LORELIA Residual Test often outperforms the common global outlier tests presented in Chapter 5. In Section 7.1.2, the superiority of the LORELIA Residual test is theoretically proven for a simple data model.

In Section 7.1.3, all outlier tests are compared on a variety of different simulated datasets which represent the most common data situations in practical applications. As the outlier tests of [Wadsworth, 1990] presented in Chapter 5 are all *global* outlier tests, the fact if the data distribution within the measuring range is homogeneous or not will not influence the performance of these tests. The *position* of existing outliers within the measuring range does not influence the performance of global outlier tests, as well. The test performance of the LORELIA Residual test however will be influenced by both criteria. Therefore the comparison is done for homogeneously distributed datasets which differ only with respect to the criteria (ii.) to (iv.). A performance ranking is given by comparing the correctness of the test results: A good outlier test should identify as many true outliers as possible (true positive test results) without wrongly declaring measurement values which belong to the population of interest as outliers (false positive test results). The true affiliation of measurement values to the population of interest P_{int} or to the contaminating population P_{cont} is usually not known for real data situations. Therefore the tests are compared on simulated datasets which contain predefined outliers.

Unlike the other outlier tests for method comparison studies presented in this work, the LORELIA Residual Test is a *local* outlier test. Thus, its performance will be influenced by the distribution of measurement values within the measuring range and by the position of existing outliers. In Section 7.2, the influence of the outlier position on the performance of the LORELIA Residual Test will be evaluated for homogeneous and inhomogeneous distributed datasets by a simulation study, compare criteria (i.) and (v.).

The LORELIA Residual Test is only appropriate if the local residual variances do not change to drastically over the measuring range and if the sample distribution is not too inhomogeneous. This problem is discussed in Section 7.3 and a solution is suggested.

The performance of the new test is influenced by the choice of the adjustment procedure for the local significance levels. Throughout this chapter, the LORELIA Residual Test was applied with respect to the Bonferroni adjusted local outlier limits. In Section 4.1.13, the choice of this adjustment procedure is discussed an an alternative method is proposed.

In Section 7.5 the results of this chapter are summarized.

All simulations in this chapter were programmed in SAS® 9.1. The underlying random number generator is based on the SAS® function RANUNI which is described in [SAS Insitute Inc., 2008]. Uniformly distributed random numbers on the interval $[0, 1]$ are generated with the pseudo random number generator proposed by [Fishman, Moore, 1982] which is given as follows:

$$x_{n+1} = (397204094 \cdot x_n) \bmod (2^{31} - 1), \text{ for a seed value } x_0 \in \left[0, 2^{31} - 1\right].$$

Other continuously distributed random numbers are generated with the help of the inverse transform sampling method.

7.1 The LORELIA Residual Test in Comparison to Common Outlier Tests

In this Section, the new LORELIA Residual Test will be compared to the outlier tests presented in Chapter 5. Thereby, remember that by (6.1.5) in Section in 6.1, the distribution of the orthogonal residuals for Passing-Bablok regression is approximately equivalent to the distribution of the measurement errors in method M_x and M_y. Therefore, it holds:

- If the residual variance is constant $\sigma_{r_i}^2 \equiv \sigma^2$, assumptions (4.1.3) and (4.1.4) of Section 4.1.1 are fulfilled and thus the absolute differences D_i^{abs} are normally distributed which corresponds to assumption (4.1.6). Therefore the global test of [Wadsworth, 1990] based on the absolute differences and the one for the orthogonal residuals are expected to deliver similar results.

- If a constant coefficient of variance is given $\sigma_{r_i}^2 = c_i^2 \cdot \sigma^2$, assumptions (4.1.12) and (4.1.13) of Section 4.1.2 are fulfilled and hence the normalized relative differences D_i^{normrel} are normally distributed which corresponds to assumption (4.1.19).

7.1.1 Performance Comparison for Real Data Situations

To give an impression on the different test performances, all tests will be compared on a variety of different exemplary datasets, which are all real data situations from clinical practice. The presented datasets differ concerning the underlying residual variance models, the data distribution within the measuring range and the number and position of suspicious outlier candidates. The evaluation of test outputs will give a first impression of the behavior and the advantages of the LORELIA Residual Test.

7.1.1.1 No Suspicious Values

The first exemplary dataset consists of $n = 147$ measurements. The distribution of data values within the measuring range is inhomogeneous. To get a visual impression of the data distribution, consider the corresponding Passing-Bablok regression plot:

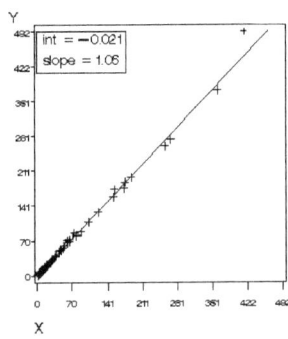

Figure 7.1: Example 1 - No Suspicious Values for Inhomogeneously Distributed Data

The majority of data is accumulated at a low concentration range. With increasing measurement values the data density decreases. The local orthogonal residual variance grows slightly with increasing measurement values (note the high magnitude of the measuring range). A visual inspection does not point out any obvious outlier candidate. The orthogonal residual at the right end of the measuring range is slightly larger than all other residuals. However, this measurement value is located in an area with low information density with only a few neighboring data points so it can not clearly be considered as an outlier candidate.

Since the residual variance increases over the measuring range, the tests of [Wadsworth, 1990] based on the absolute differences and on the orthogonal residuals deliver no appropriate results. All measurement values corresponding to a higher concentration level are identified as outliers:

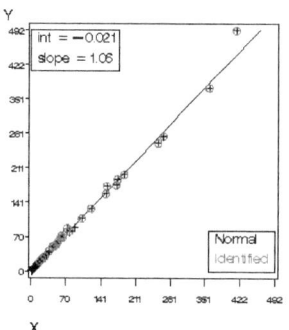 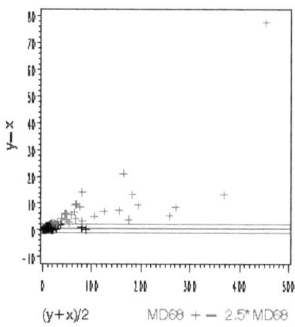

Figure 7.2: Example 1 - Outlier Test for the Absolute Differences

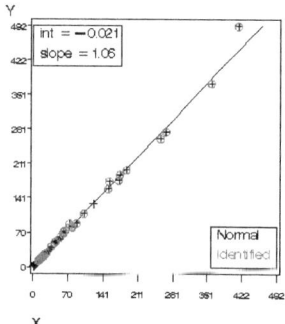 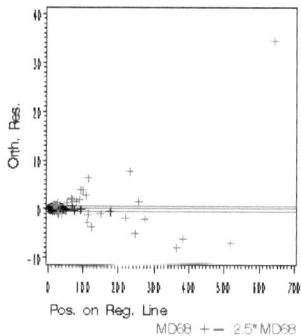

Figure 7.3: Example 1 - Outlier Test for the Residuals

The outlier test for the normalized relative differences is much more appropriate for this data situation. Note however, that two measurement values at a very low concentration level are identified as outliers, although they are not visually suspicious. The assumption that the error variances are proportional to the true concentration seems well fulfilled for higher concentrations but not appropriate at the lower concentration range. This is a common problem, which can be met in many practical examples. In [Rocke, Lorenzato, 1995], a two component error model for those data situations is proposed.

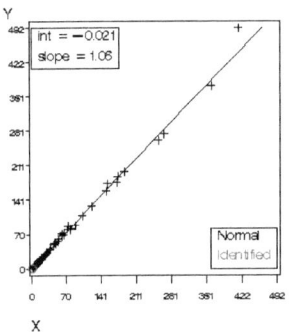
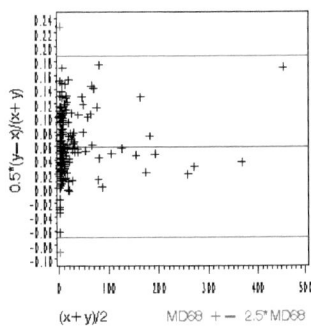

Figure 7.4: Example 1 - Outlier Test for the Normalized Relative Differences

Now, the test results obtained from the new LORELIA Residual Test are considered. The Bonferroni adjusted local significance levels α_{loc} for this exemplary dataset are given by:

$$\alpha_{\text{loc}} = \frac{\alpha_{\text{glob}}}{n} = \frac{0.1}{147} \approx 0.00068.$$

The output plots are given as follows:

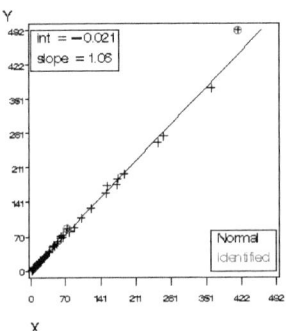
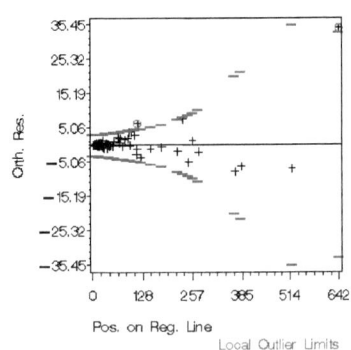

Figure 7.5: Example 1 - The LORELIA Residual Test

Two outliers are identified, which lay just slightly outside their corresponding confidence intervals. One of it correspond to the largest measurement value discussed above. Another measurement value in the middle of the measuring range lays just slightly inside its corresponding confidence interval and is thus not identified as an outlier.

The local confidence limits merge smoothly and get wider with increasing concentration. This is due to the increasing residual variance on the one hand and to the decreasing data density on the other hand as explored in Section 6.3.2.4.

It will be interesting to have a look at the reliability measures $\Gamma_{k,n}$ for all residual r_k with $k = 1, ..., n$. The identified outliers should have small reliability measures to guarantee that the residual variance estimates are not biased by the presence of the outliers. In the following, the position of (x_k^p, y_k^p) on the regression line is plotted against its corresponding reliability measure $\Gamma_{k,n}$ for $k = 1, ..., n$:

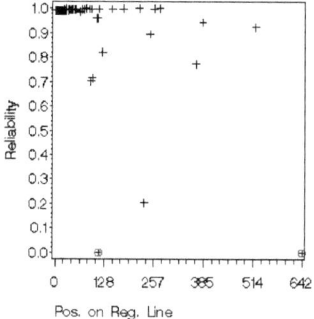

Figure 7.6: Example 1 - Reliability Plot with Identified Outliers

Both outliers have a reliability weight of 0. The value which lay just slightly inside its outlier limits is severely down weighted, as well. This value might have been detected as an outlier, if a less conservative adjustment method would have been chosen. This points out that the formal identification of outliers should always be accompanied by a visual data inspection by a data analyst experienced in the field.

Values at the low concentration range all have reliability weights near 1. For higher concentrations, several values are down weighted by a small amount. This however does not lead to the identification of many outliers. A single residual is only truly down weighted in the considered residual variance estimator, if all surrounding residual weights are higher. Therefore a low reliability weight does not necessarily correspond to a low overall weight.

7.1.1.2 One Outlier Candidate

The second exemplary dataset has a sample size of $n = 46$. The data distribution is again inhomogeneous. One obvious outlier candidate can be visually identified:

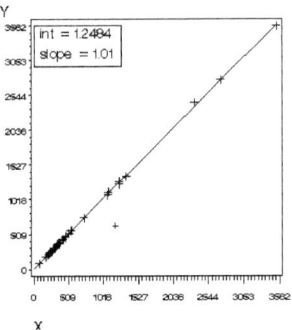

Figure 7.7: Example 2 - One Outlier Candidate for Inhomogeneously Distributed Data

The outlier test of [Wadsworth, 1990] for the absolute differences delivers the following results:

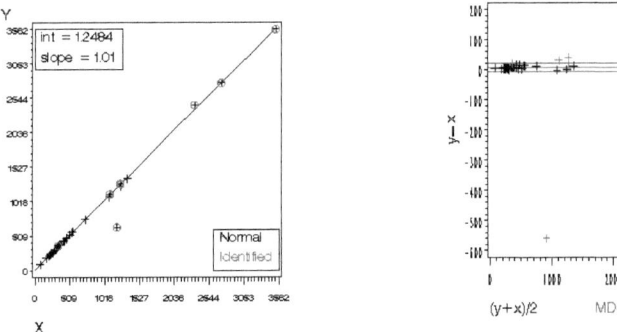

Figure 7.8: Example 2 - Outlier Test for the Absolute Differences

Again the majority of measurement values at the high concentration range is identified as outliers. The scatter plot reveals that the absolute differences are not normally distributed here. The test based on the orthogonal residual delivers similar results:

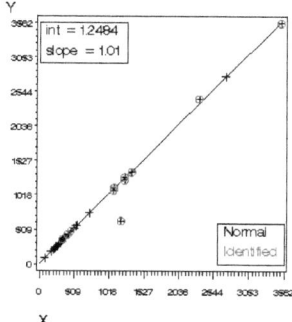 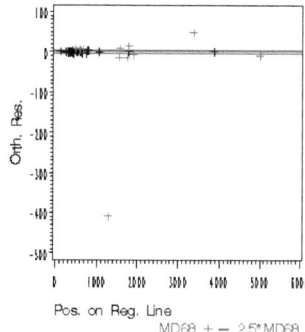

Figure 7.9: Example 2 - Outlier Test for the Residuals

The test for the normalized relative differences performs better, although the outlier limits seem to narrow as a total number of five outliers is identified. This may be due to the fact, that no adjustment of the global significance level for this multiple test situation is done.

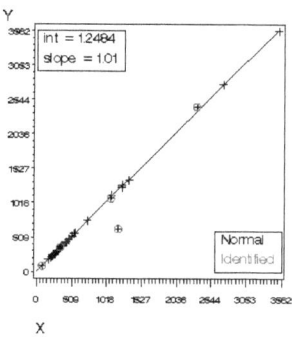 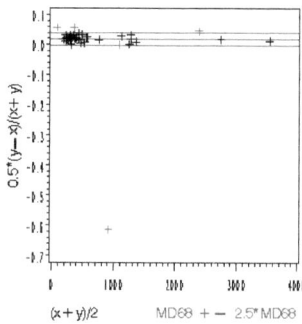

Figure 7.10: Example 2 - Outlier Test for the Normalized Relative Differences

Now, calculate the Bonferroni adjusted confidence limits for the LORELIA Residual Test:

$$\alpha_{\text{loc}} = \frac{\alpha_{\text{glob}}}{n} = \frac{0.1}{46} \approx 0.0022.$$

The test results are given as follows:

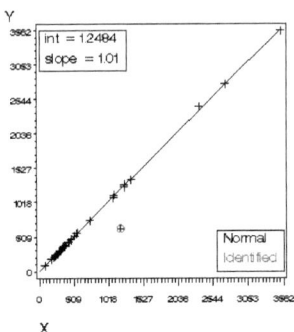 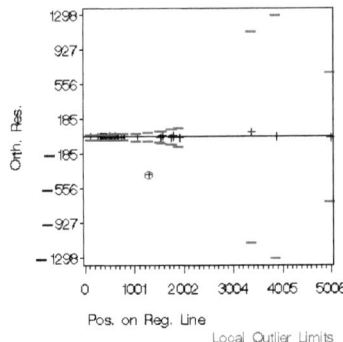

Figure 7.11: Example 2 - Identified Outliers in the Regression and the Residual Plot

One outlier is identified which correspond to the only visually suspicious value. The identified outlier lays far outside its corresponding confidence interval. Again, the outlier limits merge smoothly and get wider with increasing concentration. The confidence limits at the right limit of the measurement range are extremely wide. This is explained by the fact that data density is very low here. The reliability plot is given by:

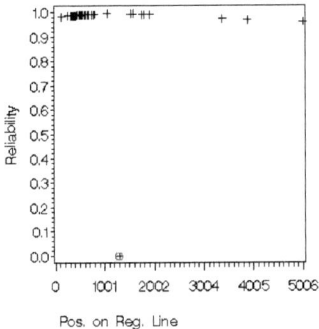

Figure 7.12: Example 2 - Reliability Plot with Identified Outlier

The identified outlier is the only one which is down weighted to an amount of 0. All other values have a local reliability measure near 1.

7.1.1.3 Uncertain Outlier Situation

The following dataset consists of $n = 42$ measurement values. The distribution of measurement values is inhomogeneous. The local residual variances increase slightly. The outlier situation is uncertain, as the most extreme residuals are located in a area with very low information density:

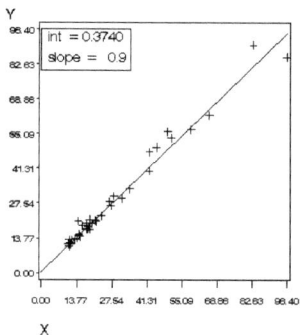

Figure 7.13: Example 3 - Uncertain Outlier Situation

The test based on the absolute differences identifies a high number of outliers. The scatter plot reveals that the normal assumption (4.1.6) is not fulfilled:

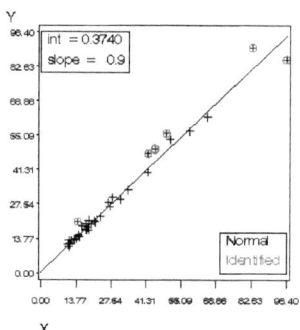 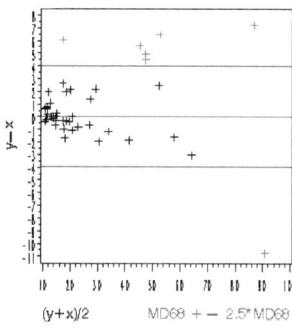

Figure 7.14: Example 3 - Outlier Test for the Absolute Differences

The test based on th orthogonal residuals delivers exactly the same results is thus not appropriate for this data situation, as well:

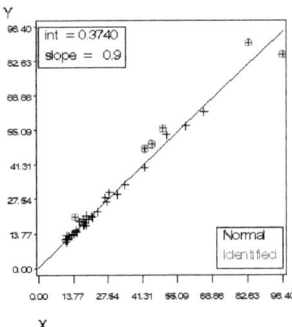
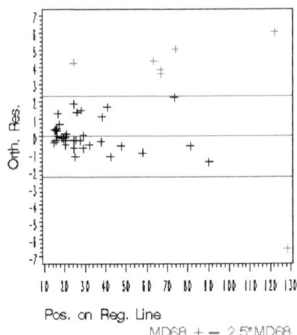

Figure 7.15: Example 3 - Outlier Test for the Residuals

The test for the normalized relative differences however is more appropriate here. One outlier is identified which is slightly larger than the surrounding residuals.

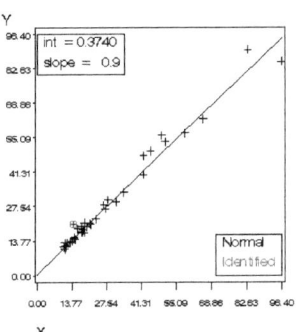
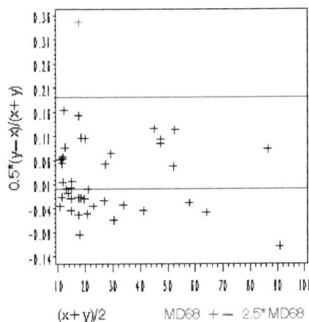

Figure 7.16: Example 3 - Outlier Test for the Normalized Relative Differences

For the LORELIA Residual Test with a local level of significance given by

$$\alpha_{\text{loc}} = \frac{\alpha_{\text{loc}}}{n} = \frac{0.1}{42} \approx 0.0024$$

no outlier at all is identified:

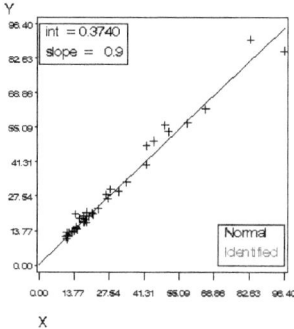 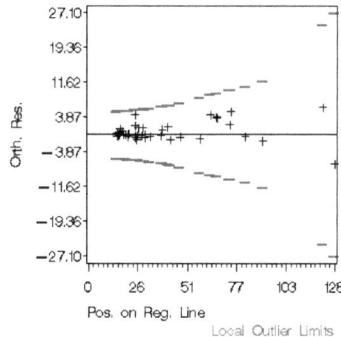

Figure 7.17: Example 3 - The LORELIA Residual Test

However, the measurement value which was identified as an outlier by the test of [Wadsworth, 1990] based on the normalized differences lays just slightly inside its corresponding confidence limits and corresponds to the value with the lowest reliability measure within the dataset:

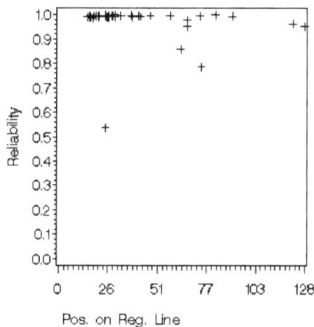

Figure 7.18: Example 3 - Reliability Plot

Remember, that the Bonferroni correction can lead to a notable loss of power. For a less conservative adjustment procedure, the LORELIA Residual Test will therefore deliver the same results as the test based on normalized relative differences.

7.1.1.4 Decreasing Residual Variances

The following example represents the unusual case of a decreasing local residual variances. Thus, none of the assumptions for the common outlier tests presented in Section 5 are fulfilled. The sample size is given by $n = 141$.

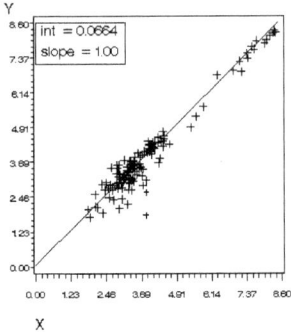

Figure 7.19: Example 4 - Decreasing Residual Variance

The global outlier test based on the absolute differences identifies to many outliers at the low concentration range:

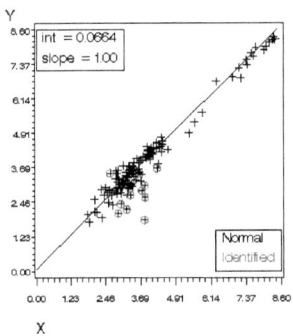 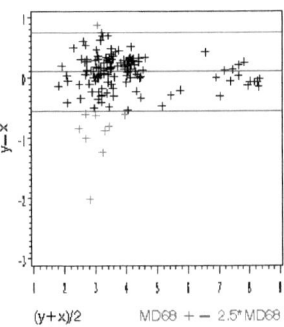

Figure 7.20: Example 4 - Outlier Test for the Absolute Differences

The test for the orthogonal residuals is not appropriate either:

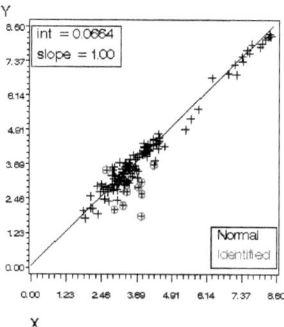 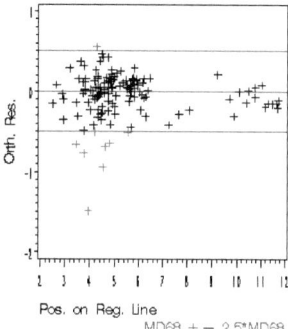

Figure 7.21: Example 4 - Outlier Test for the Residuals

The test based on the normalized relative differences performs even worse:

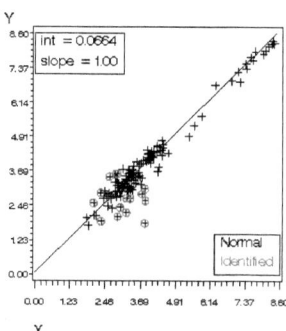 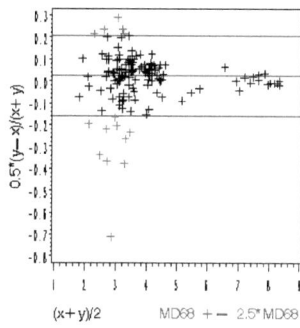

Figure 7.22: Example 4 - Outlier Test for the Normalized Relative Differences

The LORELIA Residual Test is done for a local significance level of

$$\alpha_{\text{loc}} = \frac{\alpha_{\text{loc}}}{n} = \frac{0.1}{141} \approx 0.0007.$$

Only two outliers are identified which correspond to the most extreme observations. The local outlier limits merge smoothly and model the trend of a decreasing residual variance well:

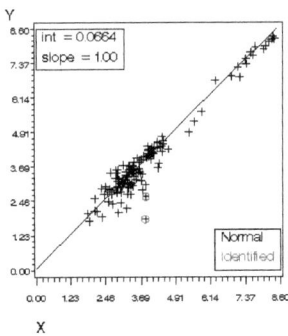 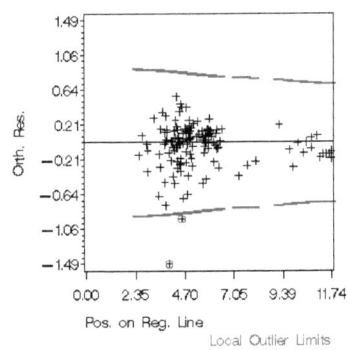

Figure 7.23: Example 4 - The LORELIA Residual Test

The reliability plot clearly shows, that the identified outliers are extremely down weighted:

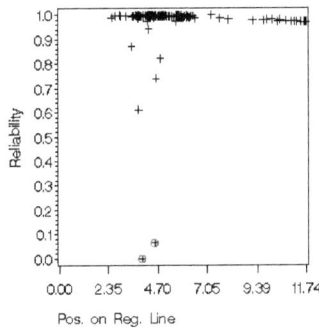

Figure 7.24: Example 4 - Reliability Plot with Identified Outliers

7.1.1.5 Very Inhomogeneous Data Distribution

The next dataset with a sample size of $n = 692$ has an extremely inhomogeneous information density. Most observations are accumulated at a low concentration level. Only a few isolated measurement values lay within the higher concentration range. One visually suspicious measurement value is located at the higher concentration range:

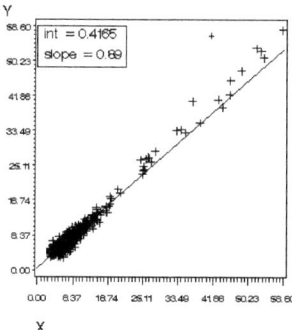

Figure 7.25: Example 5 - Very Inhomogeneous Data Dispersion

The Passing-Bablok slope estimator is mainly influenced by the cloud of low concentrated values. Measurement values corresponding to a high concentration level are nearly all located *above* the regression line. If a less robust regression method would be used like principal component analysis, the fit would be better for the high concentrated samples due to a leverage effect (compare Section 3.3.3). However in this case, most measurement values corresponding to low concentration levels would be located above the regression line. Thus, the global data trend is generally not well explained by a linear regression model.

Consider the outlier tests for the absolute differences:

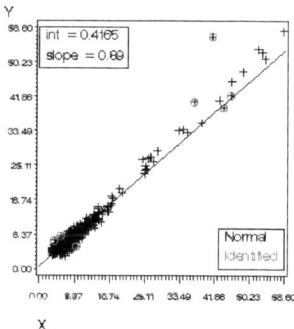
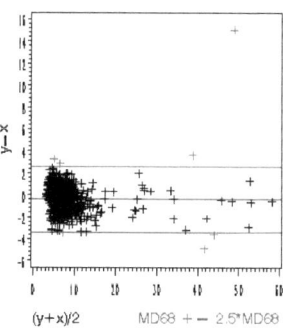

Figure 7.26: Example 5 - Outlier Test for the Absolute Differences

Six Outliers are identified here. The identified outliers do not all correspond to extreme residuals, which can be explained by the fact that the Passing-Bablok regression model is inappropriate for higher concentrations. The scatter plots can not really verify the normal assumption here as the cloud of measurement values avoid a clear visual inspection.

The test based on the orthogonal residuals identifies nearly all higher concentrated measurement values as outliers. Here, the problem with the model adjustment is even more obvious:

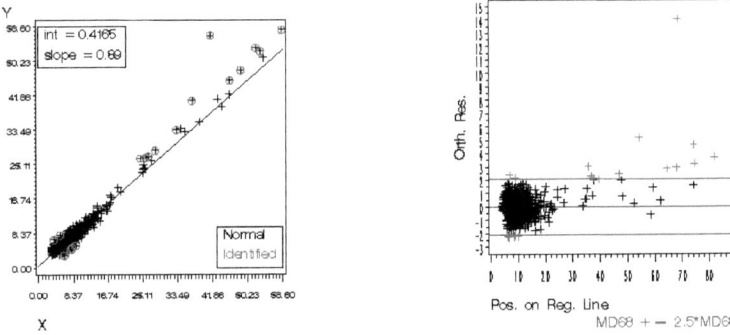

Figure 7.27: Example 5 - Outlier Test for the Residuals

The test for the normalized relative differences does not work here, as well. A huge number of outliers is identified, all located at the left limit of the measuring range. Again, the assumption of proportional increasing measurement values is especially wrong for low concentrations.

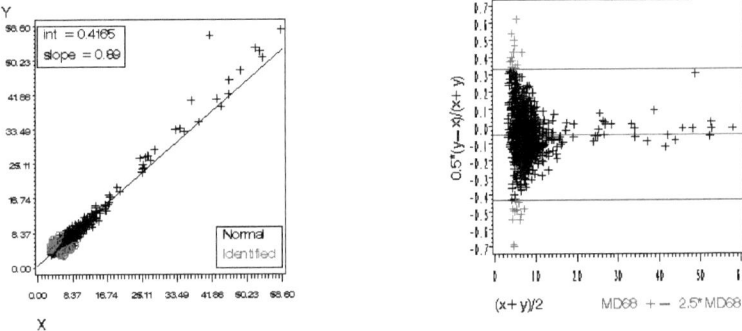

Figure 7.28: Example 5 - Outlier Test for the Normalized Relative Differences

For a local significance level of

$$\alpha_{\text{loc}} = \frac{\alpha_{\text{loc}}}{n} = \frac{0.1}{692} \approx 0.00014$$

the LORELIA Residual Test delivers the following result:

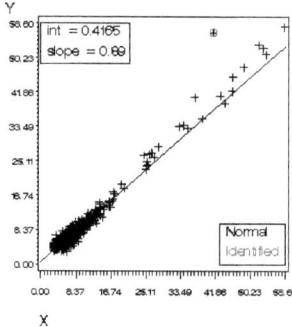 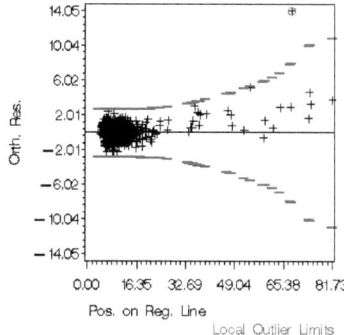

Figure 7.29: Example 5 - The LORELIA Residual Test

One outlier is identified which correspond to the visually suspicious value mentioned above. Some other residuals lay just slightly inside their corresponding confidence intervals. The reliability plot shows that several values are down weighted, but the identified outlier is the only which is down weighted to an amount of 0:

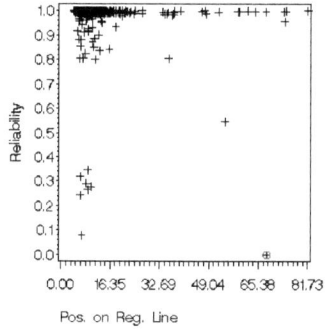

Figure 7.30: Example 5 - Reliability Plot with Identified Outliers

Note that there exists a cloud of down weighted observations at the lower concentration range. This however does not lead to the identification of a cloud of outliers, as they are all down weighted to a similar amount! Remember that a single residual is only truly down weighted, if all surrounding residual weights are higher.

7.1.1.6 Conclusion

The above real data examples represent a broad field of data situations in which the assumption for common outlier tests are not (well) fulfilled. Therefore, the outliers tests proposed by [Wadsworth, 1990] which are defined in Section 5 all deliver either wrong or misleading results. As the LORELIA Residual is based on relaxed statistical assumptions (compare Section 6.1), it clearly performs best for the presented examples. The following general observations with respect to the performance of the LORELIA Residual Test can be made:

(i.) Visually suspicious values are well identified.

(ii.) The local confidence limits merge smoothly over the measuring range.

(iii.) The LORELIA Residual Test is more conservative in areas with low information density than in areas with a high data density.

(iv.) There exist a trend of decreasing reliability when data density decreases. This does not lead to outlier misclassification, since the reliability of a residual always has to be compared to the reliabilities of the surrounding residuals.

7.1.2 Proof of Performance Superiority for an Exemplary Data Model

As pointed out above, the performance of an outlier test strongly depends on the underlying data situations. Although, the LORELIA Residual Test often outperforms the common global outlier tests proposed by [Wadsworth, 1990], the performance superiority of the new test is difficult to prove for general data models, as the expected LORELIA Weights $w_{ik} = \Delta_{ik} \cdot \Gamma_{k,n}$ depend on many different influence factors such as the sample size, the underlying residual variance model, the sample distribution within the measuring range and the number, the position and the magnitude of outliers. Moreover, the LORELIA Residual Test do not always outperform *all* of the tests presented in Chapter 5, as the test for the absolute differences, the test for the normalized relative differences and the test for the orthogonal residuals are generally not expected to deliver the same results since they are based on different statistical assumptions.

In this section, a simple model class M of method comparison datasets is defined for which all tests presented in Chapter 5 are expected to deliver the same results. This model class is defined such that the number of influence factors for the LORELIA Weights is reduced, which allows to prove the general superiority of the LORELIA Residual Test for datasets belonging to this model class. An exemplary dataset from this model class is evaluated with all outlier tests in order to illustrate the theoretical result.

The model class M is defined as follows:

Definition 7.1 (The Model Class M of Method Comparison Datasets)

A dataset S of sample size n belongs to the model class M of method comparison datasets if the following assumptions are fulfilled:

(i.) The observed measurement values of method M_x respective M_y are given as:

$$x_i = \widetilde{x}_i + \epsilon_{x_i}, \quad (7.1.1)$$
$$y_i = \widetilde{y}_i + \epsilon_{y_i}, \text{ for } i = 1, ..., n, \quad (7.1.2)$$

where $\widetilde{x}_i, \widetilde{y}_i$ are the expected measurement values and $\epsilon_{x_i}, \epsilon_{x_i}$ correspond to the random errors.

(ii.) The expected measurement values are related by a linear relationship with slope 1 and intercept 0:

$$\widetilde{x}_i = \widetilde{y}_i =: c_i, \text{ for } i = 1, ..., n, \quad (7.1.3)$$

where c_i denotes the true concentration of the ith sample. Without loss of generality it will be assumed that the order of the measurement tuples $(x_1, y_1), (x_2, y_2), ..., (x_n, y_n)$ correspond to the order of increasing true concentrations.

(iii.) The sample distribution within the measuring range is homogeneous.

(iv.) Let m be given such that $m > \frac{n}{2}$. The random errors belonging to the first m observations are given as:

$$\epsilon_{x_i} = \epsilon_{y_i} \equiv 0, \text{ for } i = 1, ..., m, \quad (7.1.4)$$

which can be equally formulated as $E_{x_i}, E_{y_i} \sim N(0, \sigma_i^2)$, with $\sigma_i^2 = 0$ for $i = 1, ..., m$. The remaining $n - m$ random errors are realizations of:

$$E_{x_i}, E_{y_i} \sim N(0, \sigma_i^2), \text{ for, } i = m+1, ..., n, \quad (7.1.5)$$

where σ_i is a multiple of the corresponding true sample concentration c_i:

$$\sigma_i = c \cdot c_i, \text{ for all } i = m+1, ..., n \text{ and a factor } c > 0. \quad (7.1.6)$$

Note that the measurement values belonging to a dataset in M are all determined by a predefined random error model. Thus, the datasets contain no true outliers due to contamination (compare Section 2.4.2).

For method comparison datasets belonging to the model class M, the following theorem can be proven:

Theorem 7.2

Let $S \in M$ be a dataset of sample size n as defined in Definition 7.1. Applying the LORELIA Residual Test and the outlier tests proposed by [Wadsworth, 1990] which are presented in Chapter 5 to the dataset S yields the following results:

(i.) *All outlier tests presented in Chapter 5 (test for the absolute differences, test for the normalized relative differences and test for the orthogonal residuals) deliver an expected number of $n - m$ false positive test results, which will correspond to the observations $(x_{m+1}, y_{m+1}), (x_{m+2}, y_{m+2}), ..., (x_n, y_n)$.*

(ii.) *All local LORELIA Residual Variance Estimates $\hat{\sigma}_1^2, \hat{\sigma}_2^2, ..., \hat{\sigma}_n^2$ will be biased. The minimal expected bias is bounded by:*

$$\xi_{\min} := E\left(\min_{\{1 \leq i \leq n\}} \{|\sigma_i^2 - \hat{\sigma}_i^2|\}\right) \leq c^2 \cdot \sum_{k=m+1}^{n} E\left(\frac{w_{1k}}{\sum_{l=1}^{n} w_{1l}}\right) \cdot c_i^2, \quad (7.1.7)$$

where the parameter c is given by (7.1.6) in Definition 7.1.

(iii.) *The maximal expected bias over all local LORELIA Residual Variance Estimates is given by:*

$$\xi_{\max} := E\left(\max_{\{1 \leq i \leq n\}} \{|\sigma_i^2 - \hat{\sigma}_i^2|\}\right) = c^2 \cdot c_n^2 - E\left(\hat{\sigma}_n^2\right). \quad (7.1.8)$$

(iv.) *For the minimal and the maximal expected bias over all local LORELIA Residual Variance Estimates it holds:*

$$\xi_{\min}, \xi_{\max} \to 0, \text{ as } c \to 0. \quad (7.1.9)$$

Proof:

(i.) By (7.1.4) and (7.1.5), the random errors of both methods are normally distributed with mean 0 and an error variance ratio of 1 where the first m observations correspond to normally distributed random errors with mean 0 and a random error variance of 0. In [Passing, Bablok, 1984] it is shown that under these conditions, the Passing-Bablok parameter estimators are unbiased. By (7.1.3), the true intercept and slope describing the linear relationship between the expected measurement values are given by 0 and 1, respectively. Therefore, it holds:

$$E(\hat{\alpha}_{PB}) = 0, \quad (7.1.10)$$
$$E(\hat{\beta}_{PB}) = 1. \quad (7.1.11)$$

From (7.1.4), (7.1.5), (7.1.10) and (7.1.11) it can be deduced that the random variables for the absolute differences, for the normalized relative differences and for the orthogonal residuals fulfill:

$$\left.\begin{array}{r} \mathrm{E}(D_i^{\mathrm{abs}}) \\ \mathrm{E}(D_i^{\mathrm{normrel}}) \\ \mathrm{E}(R_i) \end{array}\right\} = 0, \text{ for } i = 1, ..., n,$$

where D_i^{abs} and D_i^{normrel} are defined trough (4.1.5) and (4.1.18) in Chapter 4. As the median is an unbiased estimator for the mean of normally distributed random variables, it follows:

$$\left.\begin{array}{r} \mathrm{E}(\mathrm{med}(D^{\mathrm{abs}})) \\ \mathrm{E}(\mathrm{med}(D^{\mathrm{normrel}})) \\ \mathrm{E}(\mathrm{med}(R)) \end{array}\right\} = 0.$$

By (7.1.4) and (7.1.5), the expected order of the absolute deviations between the absolute differences and their median is given by:

$$\begin{aligned} |D_1^{\mathrm{abs}} - \mathrm{med}(D^{\mathrm{abs}})| &\leq |D_2^{\mathrm{abs}} - \mathrm{med}(D^{\mathrm{abs}})| \\ &\leq ... \\ &\leq |D_m^{\mathrm{abs}} - \mathrm{med}(D^{\mathrm{abs}})| \\ &< |D_{m+1}^{\mathrm{abs}} - \mathrm{med}(D^{\mathrm{abs}})| \\ &< ... \\ &< |D_n^{\mathrm{abs}} - \mathrm{med}(D^{\mathrm{abs}})|. \end{aligned}$$

Thus by (ii.) in Definition 7.1, the expected order correspond to the order of increasing concentrations c_i. The same holds true for the expected order of the absolute deviations between the normalized relative differences and their median or for the orthogonal residuals.

By [Hartung et al., 2009] (Chaper XIV, Page 856), the 68% median absolute deviation which is given as 1.4828 times the median absolute deviation, is an unbiased estimator for the standard deviation of a normal distribution. By the above considerations, the expected ranks of the absolute deviations between the respective comparison measures and their median are expected to correspond to the order of the measurement values. Thus, as $m > \frac{n}{2}$, it holds by (7.1.4):

$$
\begin{aligned}
&\mathrm{E}\left(\mathrm{med}(|D^{\mathrm{abs}} - \mathrm{med}(D^{\mathrm{abs}})|)\right) \\
&= \begin{cases} \mathrm{E}\left(|D^{\mathrm{abs}} - \mathrm{med}(D^{\mathrm{abs}})|_{(\frac{n+1}{2})}\right), & \text{if } n \text{ is odd,} \\ \mathrm{E}\left(0.5 \cdot \left[|D^{\mathrm{abs}} - \mathrm{med}(D^{\mathrm{abs}})|_{(\frac{n}{2})} + |D^{\mathrm{abs}} - \mathrm{med}(D^{\mathrm{abs}})|_{(\frac{n+2}{2})}\right]\right), & \text{if } n \text{ is equal,} \end{cases} \\
&= \begin{cases} \mathrm{E}\left(|D^{\mathrm{abs}} - \mathrm{med}(D^{\mathrm{abs}})|_{\frac{n+1}{2}}\right), & \text{if } n \text{ is odd,} \\ \mathrm{E}\left(0.5 \cdot \left[|D^{\mathrm{abs}} - \mathrm{med}(D^{\mathrm{abs}})|_{\frac{n}{2}} + |D^{\mathrm{abs}} - \mathrm{med}(D^{\mathrm{abs}})|_{\frac{n+2}{2}}\right]\right), & \text{if } n \text{ is equal,} \end{cases} \\
&= 0.
\end{aligned}
$$

Equivalently, on may show that:

$$
\begin{aligned}
\mathrm{E}\left(\mathrm{med}(|D^{\mathrm{normrel}} - \mathrm{med}(D^{\mathrm{normrel}})|)\right) &= 0, \\
\mathrm{E}\left(\mathrm{med}(|R - \mathrm{med}(R)|)\right) &= 0.
\end{aligned}
$$

Hence, it holds:

$$
\left.\begin{aligned}
\mathrm{E}\left(\mathrm{mad68}(D^{\mathrm{abs}})\right) \\
\mathrm{E}\left(\mathrm{mad68}(D^{\mathrm{normrel}})\right) \\
\mathrm{E}\left(\mathrm{mad68}(R)\right)
\end{aligned}\right\} = 0.
$$

Therefore, the expected global outlier limits defined by (5.1.1) and (5.2.2) in Section 5 are given by:

$$
\left.\begin{aligned}
\mathrm{E}\left(\mathrm{med}(d^{\mathrm{abs}}) \pm 2.5 \cdot \mathrm{mad68}(d^{\mathrm{abs}})\right) \\
\mathrm{E}\left(\mathrm{med}(d^{\mathrm{normrel}}) \pm 2.5 \cdot \mathrm{mad68}(d^{\mathrm{normrel}})\right) \\
\mathrm{E}\left(\mathrm{med}(r) \pm 2.5 \cdot \mathrm{mad68}(r)\right)
\end{aligned}\right\} = \pm 0.
$$

Thus, all measurement values which correspond to a residual variance $\sigma_i^2 > 0$, namely observations $(x_{m+1}, y_{m+1}), (x_{m+2}, y_{m+2}), \ldots, (x_n, y_n)$, are expected to be wrongly identified as outliers.

$\square_{(\mathrm{i}.)}$

(ii.) The LORELIA Weights do not involve any model information of the underlying residual variances in order to be globally applicable for every data situation. As the local residual variances are not constant over the whole measuring range for datasets belonging to the model class M, all local LORELIA Residual Variance Estimates will be biased due to a smoothing effect. This bias will be minimal for the first measurement value (x_1, y_1) as the nearest $m - 1$ neighbors correspond to the same local residual variance of 0. It is only slightly overestimated due to the remaining $n - m$ measurement values which lay far away from (x_1, y_1) and which correspond to residual variances greater than 0. Thus, it holds:

$$
\xi_{\min} := \mathrm{E}\left(\min_{\{1 \leq i \leq n\}} \left\{|\sigma_i^2 - \widehat{\sigma}_i^2|\right\}\right) = \mathrm{E}\left(\widehat{\sigma}_1^2\right).
$$

An upper bound for expected value of the 1st LORELIA Residual Variance Estimate $\hat{\sigma}_1^2$ is calculated in the following. Thereby, remember that by Definition (6.5) of the LORELIA Reliability Measure $\Gamma_{k,n}$ in Section 6.4.2, the values of the LORELIA Weights w_{ik} are statistically dependent of the random variables R_k^2.

$$\begin{aligned}
E\left(\hat{\sigma}_1^2\right) &= E\left(\frac{1}{\sum_{l=1}^n w_{1l}} \cdot \sum_{k=1}^n w_{1k} \cdot R_k^2\right) \\
&= E\left(\frac{1}{\sum_{l=1}^n w_{1l}} \cdot \sum_{k=1}^m w_{1k} \cdot r_k^2\right) + E\left(\frac{1}{\sum_{l=1}^n w_{1l}} \cdot \sum_{k=m+1}^n w_{1k} \cdot R_k^2\right) \\
&\leq \sum_{k=1}^m E\left(\frac{w_{1k}}{\sum_{l=1}^n w_{1l}}\right) \underbrace{E\left(r_k^2\right)}_{=0 \text{ by } (7.1.4)} + \sum_{k=m+1}^n E\left(\frac{w_{1k}}{\sum_{l=1}^n w_{1l}}\right) \underbrace{E\left(R_k^2\right)}_{=c^2 \cdot c_k^2 \text{ by } (7.1.5)} \\
&= c^2 \cdot \sum_{k=m+1}^n E\left(\frac{w_{1k}}{\sum_{l=1}^n w_{1l}}\right) \cdot c_k^2.
\end{aligned}$$

This proves (ii.).

$\square_{(ii.)}$

(iii.) The LORELIA Residual Variance Estimate of the limiting value (x_n, y_n) is underestimated as all neighbored residuals correspond to smaller local residual variances:

$$\begin{aligned}
0 \leq E\left(\hat{\sigma}_n^2\right) &= E\left(\frac{1}{\sum_{l=1}^n w_{nl}} \cdot \sum_{k=1}^n w_{nk} \cdot R_k^2\right) \\
&= E\left(\frac{1}{\sum_{l=1}^n w_{nl}} \cdot \sum_{k=1}^m w_{nk} \cdot r_k^2\right) + E\left(\frac{1}{\sum_{l=1}^n w_{nl}} \cdot \sum_{k=m+1}^n w_{nk} \cdot R_k^2\right) \\
&\leq \sum_{k=1}^m E\left(\frac{w_{nk}}{\sum_{l=1}^n w_{nl}}\right) \underbrace{E\left(R_k^2\right)}_{=0 \text{ by } (7.1.4)} + \sum_{k=m+1}^n E\left(\frac{w_{nk}}{\sum_{l=1}^n w_{nl}}\right) \underbrace{E\left(R_k^2\right)}_{=c^2 \cdot c_i^2 \text{ by } (7.1.5)} \\
&= c^2 \cdot \sum_{k=m+i}^n E\left(\frac{w_{ik}}{\sum_{l=1}^n w_{il}}\right) \cdot c_i^2 \\
&\leq c^2 \cdot c_n^2 \cdot \underbrace{\sum_{k=m+i}^n E\left(\frac{w_{ik}}{\sum_{l=1}^n w_{il}}\right)}_{\leq 1} \\
&\leq c^2 \cdot c_n^2.
\end{aligned}$$

The value (x_n, y_n) also correspond to the maximal bias over all local LORELIA Residual Variance Estimates as the sample distribution of S is homogeneous and the distance measures Δ_{ik} between two neighbored residuals are thus expected to be equal. Therefore, it holds:

$$\begin{aligned}
\xi_{\max} &:= \mathrm{E}\left(\max_{\{1\leq i\leq n\}}\{|\sigma_i^2 - \widehat{\sigma}_i^2|\}\right) \\
&= \mathrm{E}\left(|\sigma_n^2 - \widehat{\sigma}_n^2|\right) \\
&= c^2 \cdot c_n^2 - \mathrm{E}\left(\widehat{\sigma}_n^2\right) > 0. \quad (7.1.12)
\end{aligned}$$

This proves (iii.).

$\square_{\text{(iii.)}}$

(iv.) The proof of (iv.) follows directly from (ii.), (iii.) and (7.1.12).

$\square_{\text{(iv.)}}$

This completes the proof of Theorem 7.2.

$\square$

Remark 7.3

For datasets $S \in M$, the local LORELIA Residual Variances will generally be overestimated at the low concentration range and underestimated for higher concentrations. Therefore, if (x_n, y_n) is not falsely identified as an outlier, then there will be no false positive test results at all. By Theorem 7.2 (iv.), the increment of the bias of $\widehat{\sigma}_n^2$ is controlled by the magnitude of the parameter c, defined in (7.1.6). However, the LORELIA Outlier Limits do not exclusively depend on the residual variance estimate but also on the local adjusted significance level α_{loc}. A conservative adjustment of the global significance level such as the common Bonferroni adjustment will balance the downward bias of $\widehat{\sigma}_n^2$ to a certain extend.

For illustration, consider the following exemplary dataset belonging to the model class M:

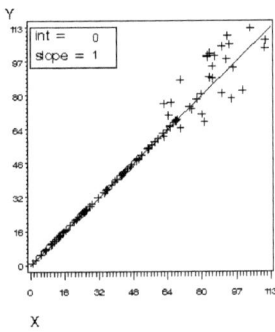

Figure 7.31: Exemplary Dataset from the Model Class M

The outlier tests proposed by [Wadsworth, 1990] deliver exactly the expected results shown in Theorem 7.2 - all values corresponding to a residual variance greater than 0 are wrongly identified as outliers:

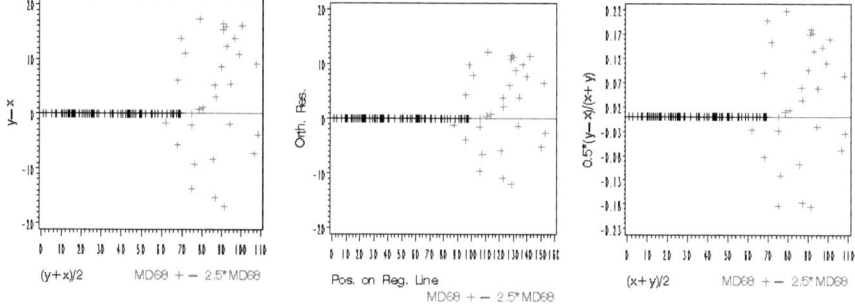

Figure 7.32: Evaluation of the Exemplary Dataset with the Global Outlier Tests Based on the Absolute Differences, on the Orthogonal Residuals and on the Normalized Relative Differences

The LORELIA Residual Test however delivers no false test results:

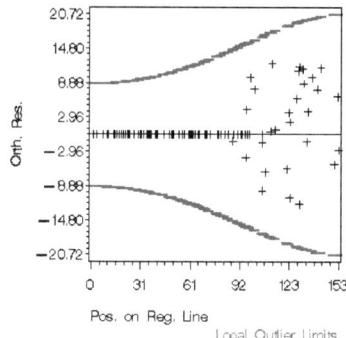

Figure 7.33: Evaluation of the Exemplary Dataset with the LORELIA Residual Test

7.1.3 Performance Comparison for Simulated Datasets

In this section, all tests will be compared on a variety of simulated datasets, which contain some known simulated outliers. The outlier tests will be compared concerning the correctness of the test results, so to say the number of true positive and false positive test results. Thereby, the comparison is done for different data situations to judge the influence of the underlying residual variance model, the number of present outliers, the magnitude of the outlier terms and the influence of the outlier distribution within the dataset.

7.1.3.1 Simulation Models

It is obvious that the test results do not uniquely depend on the choice of the outlier test but also on the actual data situation. Therefore, several representative simulation models will be introduced in this section which differ with respect to the criteria mentioned above. The general notations for all data simulations have been introduced in Section 4. The special distribution and parameter setting for the different simulation models are specified in the following. The choice of these settings is to a certain extend arbitrary. It is motivated in intention to simulate datasets which correspond well to common outlier scenarios in method comparison studies. In the context of this work, the author examined a broad range of different experimental datasets, which lead to the choice of the following simulation models.

Note that, unlike the LORELIA Residual Test, the outlier tests presented in Chapter 5 are all *global* outlier tests which will not react to the local data density, which is a clear drawback for an appropriate data analysis. The LORELIA Residual Test is more conservative in areas with low data density than in dense data areas, whereas all other tests deliver their test results independently of the local data density. Therefore, an outlier within an area of low local data density may not be identified by the LOELIA Residual Test, which is hence a false negative test result, whereas the outlier is well identified by a respective global test. However, the test result of the LORELIA Residual Test is still more appropriate as it takes the local information density into account. If the distribution of measurement values within the measuring range is inhomogeneous, the test results of the LORELIA Residual Test can therefore not directly be compared to the test results of the other tests. For this reason, the number of correct test results is only an appropriate measure for a performance comparison in the case of a homogeneous data distribution.

The different simulation models are given as follows. Consider a general sample size of:

$$n = 100. \tag{7.1.13}$$

The expected measurement values of method M_x and M_y are assumed to be equal to the true sample concentration:

$$\widetilde{x}_i = \widetilde{y}_i = c_i, \quad \text{for } i = 1, ..., n, \tag{7.1.14}$$

which correspond to the case of equivalent methods M_x and M_y.

The true concentrations c_i are homogeneously distributed within the measuring range:

$$C_i \sim U(0, 100), \quad i=1,...,n. \tag{7.1.15}$$

The measurement values including the random errors are thus realizations of:

$$X_i \sim c_i + N(0, \sigma_{r_i}^2) \tag{7.1.16}$$
$$Y_i \sim c_i + N(0, \sigma_{r_i}^2), \quad \text{for } i = 1, ..., n. \tag{7.1.17}$$

Three different residual variance models are considered.

(i.) The most simplest case of a constant residual variance will be modeled by:

$$\sigma_{r_i}^2 \equiv 0.1, \quad \text{for all } i = 1, ..., n. \tag{7.1.18}$$

(ii.) The case of a constant coefficient of variance as introduced in (4.1.14) in Section 4.1.2 is given by:

$$\sigma_{r_i}^2 = 0.01 \cdot c_i^2, \quad \text{for } i = 1, ..., n. \tag{7.1.19}$$

(iii.) The case of a non constant coefficient of variance is modeled as:

$$\sigma_{r_i}^2 = 4 + 0.01 \cdot c_i^2, \quad \text{for } i = 1, ..., n. \tag{7.1.20}$$

Outliers will be biased realizations of X_i and/or Y_i. They will be modeled as:

$$X_i + out_{x_i} \sim Z + E_{x_i} + out_{x_i} \tag{7.1.21}$$
$$Y_i + out_{y_i} \sim \tilde{x}_i + E_{y_i} + out_{y_i}, \quad \text{for } out_{x_i}, out_{y_i} \in \mathbb{R}, \ i = 1, ..., n. \tag{7.1.22}$$

If for an $i = 1, ..., n$, it holds that

$$out_{x_i} = out_{y_i} \neq 0, \tag{7.1.23}$$

this correspond to a problem within the ith sample, like a false concentration or a somehow contaminated sample. If

$$out_{x_i} \neq out_{y_i} \text{ and } (out_{x_i} = 0 \text{ or } out_{y_i} = 0), \tag{7.1.24}$$

then there was a problem in the measurement process of method M_x or method M_y respectively. In this work, only the second case (7.1.24) is considered, since an equal error term in both methods is almost impossible to detect. Without loss of generality, it will be assumed in the following that if the ith value is an outlier, this correspond to $out_{x_i} \neq 0$ and $out_{y_i} = 0$.

The outlier term is given as a multiple of the corresponding standard deviation of the underlying local residual variance:

$$out_{x_i} = k \cdot \sigma_{r_i}, \text{ for a constant } k > 1. \tag{7.1.25}$$

Two different magnitudes for the outlier term are considered:

(i.) A medium outlier term is given by:

$$out_{x_i} = 4 \cdot \sigma_{r_i}, \tag{7.1.26}$$

(ii.) A high outlier term is modeled as:

$$out_{x_i} = 8 \cdot \sigma_{r_i}. \tag{7.1.27}$$

The number of simulated outliers is given by $0, 1$ or 3.

The positions of the simulated outliers within the dataset are determined by the ranks of the observed x-values. Let $x_{(1)}, ..., x_{(n)}$ be the ordered sequence of the observed measurement values $x_1, ..., x_n$. Inf a single outlier is simulated, the outlier term is added to $x_{(50)}$, so the outlier is situated in the middle of the measuring range:

$$out_{x_i} = out_{x_{(50)}}. \tag{7.1.28}$$

If 3 uniformly distributed outliers are simulated, the outliers will have the following positions:

$$out_{x_{i_1}}, out_{x_{i_2}}, out_{x_{i_3}} = out_{(25)}, out_{(50)}, out_{(75)}. \tag{7.1.29}$$

In the case of 3 clustered outliers, the outlier positions are given es follows:

$$out_{x_{i_1}}, out_{x_{i_2}}, out_{x_{i_3}} = out_{(49)}, out_{(50)}, out_{(51)}. \tag{7.1.30}$$

The following table will give an overview of the resulting 21 data situations. For each data situation, a total number of 100 datasets is simulated.

Residual Variance	Outlier...		
	Number	Magnitude	Distribution
Constant	0		
	1	Medium	
		High	
	3	Medium	Uniformly
			Clustered
		High	Uniformly
			Clustered
Constant CV	0		
	1	Medium	
		High	
	3	Medium	Uniformly
			Clustered
		High	Uniformly
			Clustered
Non Constant CV	0		
	1	Medium	
		High	
	3	Medium	Uniformly
			Clustered
		High	Uniformly
			Clustered

Table 7.1: Considered Data Situations for the Outlier Tests Comparison

7.1.3.2 Evaluation of the Simulation Results

The $21 \cdot 100$ simulated datasets are evaluated with the new LORELIA Residual Test and with the common outlier tests presented in Chapter 5. The test results are compared with respect to the number of true positive (tp) and false positive (fp) test results and with respect to their actual type 1 error rate. Thereby, the number of false positive test results should not be confounded with the global type 1 error rate of the outlier test. The type 1 error of an outlier test corresponding to the hypotheses formulated in (3.1.1) in Section 3 is given by the probability to identify at least one outlier, when in fact there are none, independently of the fact *how many* false positive outliers are identified.

Beside the tabulated numbers of true positive and false positive test results, a visual inspection of the corresponding plots (regression plot and residual plot/scatter plot) can give important supplementary information on the test properties and their behavior for different data situations. The total number of resulting plots however can not all be shown here. Therefore, Appendix B shows the plots of one representative dataset for each of the 21 considered data situations.

Actual Type 1 Error Rates

In Chapter 5, it has already been discussed that the outlier tests proposed by [Wadsworth, 1990] do not include an adjustment of the global significance level, whereas the new LORELIA Residual Test is used with the common Bonferroni correction throughout this chapter. Thus, before the numbers of true positive and false positive test results are compared, it will be interesting to have a look at the actual type 1 error rates of the different tests. The following table list the different type 1 error rates which are approximated separately for all three residual variance models based on the datasets containing no simulated outliers:

Residual Variance	**Type 1 Error Rates**			
	Absolute Differences	Orth. Residuals	Norm. Rel. Differences	LORELIA Res. Test
Constant	0.86	0.81	1	0.13
Constant CV	1	1	0.67	0.26
Non Constant CV	1	1	1	0.23

Table 7.2: Approximated Type 1 Error Rates

It can clearly be seen that the global tests of [Wadsworth, 1990] do not meet any reasonable significance level. If the statistical assumptions on the underlying error variance model are not fulfilled, the approximated type 1 error rates are 100%, which means that for each of the 100 underlying dataset at least one false positive outlier was identified. However, even if the statistical assumptions are met, the type 1 error rates range between 67% and 86%.

The LORELIA Residual Test performs much better. If a constant residual variance is simulated, the type 1 error rate approximately meets the global significance level of 10%. Thus, also the Bonferroni adjustment procedure is expected to be very conservative, the acutal type 1 error rate is approximately equal to the global significance level.

For non constant residual variances, the local residual variance estimates will be biased due to a smoothing effect. For the underlying simulation models, this effect causes the increment in the type 1 error rates to 23% and 26%, respectively. However, the rates may be different for other simulation models, as the magnitude of the bias in the residual variance estimates directly effects the type 1 error rate.

Thus for the considered simulation models, the LORELIA Residual Test generally correspond to the smaller type 1 error rates than the tests proposed by [Wadsworth, 1990]. The global significance level α_{glob} is approximately met if the underlying residual variances are constant.

True Positive and False Positive Test Results

The following tables list the means of the true positive and false positive test results over the 100 simulated datasets for all considered data situations. Always remember that the common global outlier test do not include an adjustment for the multiple test situation, so the new LORELIA Residual Test is expected to be more conservative in all data situations. To begin with, consider the test results in case of constant residual variance:

$\sigma_{r_i}^2 \equiv 0.1$			Absolute Differences	Orth. Residuals	Norm. Rel. Differences	LORELIA Res. Test
0 Outliers			2.59 fp	2.64 fp	14.75 fp	0.15 fp
1 Outlier	Medium		0.67 tp	0.67 tp	0.56 tp	0.3 tp
			2.38 fp	2.41 fp	14.48 fp	0.11 fp
	High		1 tp	1 tp	1 tp	1 tp
			2.38 fp	2.4 fp	14.48 fp	0.03 fp
3 Outliers	Medium	Uniform	2.1 tp	2.1 tp	1.7 tp	0.79 tp
			2 fp	2.1 fp	13.83 fp	0.03 fp
		Clustered	2.04 tp	2.06 tp	1.71 tp	0.74 tp
			1.9 fp	2.1 fp	13.99 fp	0.03 fp
	High	Uniform	3 tp	3 tp	2.93 tp	2.93 tp
			1.99 fp	2.08 fp	13.81 fp	0 fp
		Clustered	3 tp	3 tp	2.99 tp	2.94 tp
			1.9 fp	2.1 fp	13.99 fp	0 fp

Table 7.3: Means of True Positive and False Positive Test Results - Homogeneous Data Distribution, Constant Residual Variance

In order to simplify the comparison of the different test result, the following bar diagrams will show the percentages of true positive test results (with respect to the total number of simulated outliers) and the percentages of false positive test results (with respect to the total number of values belonging to the population of interest P_{int}).

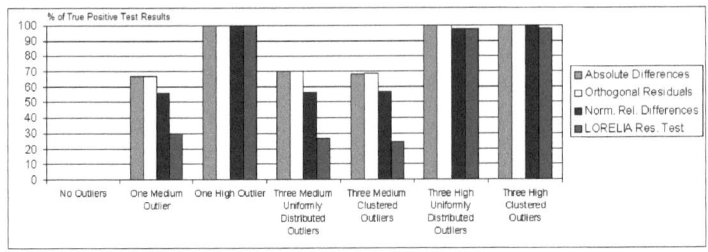

Figure 7.34: Percentages of True Positive Test Results, Constant Residual Variance

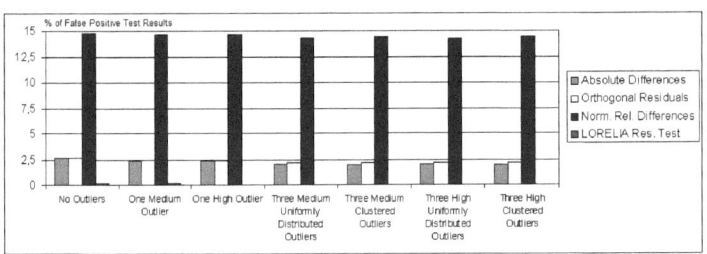

Figure 7.35: Percentages of False Positive Test Results, Constant Residual Variance

The following observations can be made:

(i.) A medium outlier term is generally not well identified. A look at the corresponding plots in Appendix B shows, that a medium outlier is often hidden in the main body of the data and thus no outlier test will be able to separate this outlier from the normal data. Therefore, it seems more appropriate to judge the outlier test performances for a higher outlier term.

(ii.) The fact if the outliers are clustered or uniformly distributed over the measuring range has no observable influence on the test performances.

(iii.) The performances for the outlier tests based on the absolute differences and on the orthogonal residuals are almost identical. This is not astonishing since these tests are based on the assumption of constant residual variances or constant error variances, respectively, which is approximately equivalent (compare (6.1.5) in Section 6.1). Both tests deliver appropriate results. The percentages of false positive test results are approximately constant for every data situation with about 2.5%.

(iv.) The test based on the normalized relative differences is not appropriate in the case of a constant residual variances. The test delivers too many false positive test results which are all located in the low concentration range (compare plots in Appendix B).

(v.) The LORELIA Residual Test is more conservative than the test of [Wadsworth, 1990] based on the absolute differences or on the orthogonal residuals. Remember, that the local confidence limits for the LORELIA Residual Test are calculated for a Bonferroni adjusted significance level of $\frac{\alpha_{glob}}{n} = \frac{0.1}{100} = 0.1\%$, whereas the tests of [Wadsworth, 1990] are not assigned to a predefined significance level but correspond to very high type 1 error rates, compare Table 7.2. Thus, the level of significances of the different outlier test are not equal. This has to be kept in mind in the comparison of the tests.

For the LORELIA Residual Test, the percentages of true positive test results are especially low for medium outliers. This can however be explained by (i.). For a high outlier term however, the percentages of true positive test results are nearly 100% whereas the percentages of false positive test remain very small.

Generally, the LORELIA Residual Test clearly delivers the best results for the high outlier term, as it separates these outliers best. The test performances for a medium outlier term have to be judged with care, as a medium outlier may not always be extreme with respect to the main body of the data.

Now consider the test results an the corresponding bar diagrams in case of a homogeneous data distribution and a constant coefficient of variance:

$\sigma_{r_i}^2 = 0.01 \cdot c_i^2$			Absolute Differences	Orth. Residuals	Norm. Rel. Differences	LORELIA Res. Test
0 Outliers			10.36 fp	10.84 fp	1.56 fp	0.3 fp
1 Outlier	Medium		0.86 tp	0.85 tp	0.44 tp	0.09 tp
			10 fp	10.4 fp	1.53 fp	0.25 fp
	High		1 tp	1 tp	0.98 tp	0.93 tp
			10 fp	10.39 fp	1.52 fp	0.22 fp
3 Outliers	Medium	Uniform	1.88 tp	1.88 tp	1.24 tp	0.49 tp
			9.22 fp	9.37 fp	1.29 fp	0.21 fp
		Clustered	2.52 tp	2.49 tp	1.12 tp	0.15 tp
			9.35 fp	9.76 fp	1.39 fp	0.2 fp
	High	Uniform	2.93 tp	2.92 tp	2.91 tp	1.81 tp
			9.18 fp	9.27 fp	1.26 fp	0.09 fp
		Clustered	3 tp	3 tp	2.92 tp	2.47 tp
			9.31 fp	9.53 fp	1.37 fp	0.06 fp

Table 7.4: Means of True Positive and False Positive Test Results - Homogeneous Data Distribution, Constant Coefficient of Variance

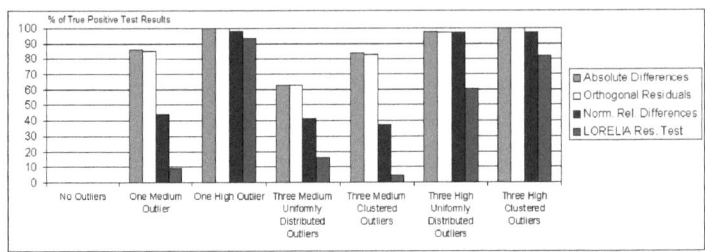

Figure 7.36: Percentages of True Positive Test Results for a Constant Coefficient of Variance

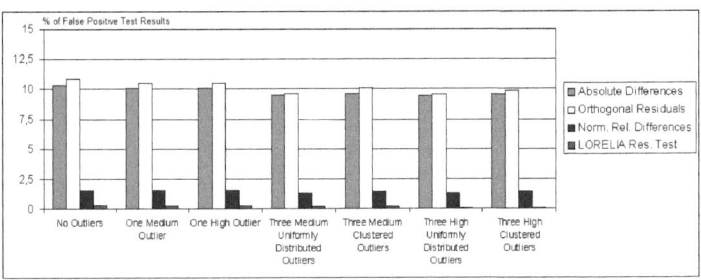

Figure 7.37: Percentages of False Positive Test Results for a Constant Coefficient of Variance

The above bar diagrams allow the following conclusions:

(i.) The medium outlier term is again not well identified.

(ii.) The outlier tests for the absolute differences and for the orthogonal residuals both deliver too many false positive test results which are all located at the end of the measuring range (compare plots in Appendix B) which is explained by the fact that the residual variances are proportionally increasing for this data model. Although the percentages of true positive test results are rather high, both test are not appropriate here as the percentages of false positive test results are of unacceptable magnitude with about 10% for all data situations.

(iii.) The test based on the normalized relative differences is the one of choice in the case of a constant coefficient of variance. A high outlier term is always well identified. The percentage of false positive test results is approximately constant for all data situations with about 1.4%.

(iv.) The LORELIA Residual Test is clearly more conservative than the test based on the normalized relative differences. The percentages of true positive test results are large in the case of one high outlier (93%) and in the case of three high clustered outliers (82, 33%). As the percentages of false positive test results are very low ($\leq 0.3\%$), the LORELIA Residual Test can be regarded as a more conservative alternative to the test for the normalized relative differences in these data situations.

(v.) The fact if the outliers are clustered or not seems to influence the percentages of true positive test results for the LORELIA Residual Test. This first seems astonishing, since there was no observable influence in the case of a constant residual variance. However this contradiction can easily be explained. The percentages of true positive test results are not really influenced by the outlier cluster but by the *position* of the outliers. The LORELIA Weighting Function does not involve any model information of the underlying residual variances in order to be globally applicable for every data situation. Therefore, in the case of a non constant residual variance, the local variance estimates will be smoothed, especially if the differences between the local residual variances are large. This means that the local LORELIA Outlier Limits will represent the trend of the underlying residual variance model, but the local variance estimators will be biased. The increment of this bias depend on the actual data situation. In the case of a constant coefficient of variance, the local residual variance will be overestimated at the low concentration range and underestimated for higher concentrations. Therefore, outliers at the low concentration range are less easily detected than at a higher concentration level. A detailed description of the influence of the outlier position for different data situations will be given in Section 7.2.

Now, consider the case of a non constant coefficient of variance.

$\sigma_{r_i}^2 = 4 + 0.01 \cdot c_i^2$			Absolute Differences	Orth. Residuals	Norm. Rel. Differences	LORELIA Res. Test
0 Outliers			7.21 fp	7.25 fp	6.49 fp	0.27 fp
1 Outlier	Medium		0.75 tp	0.77 tp	0.23 tp	0.07 tp
			6.97 fp	7.09 fp	6.29 fp	0.24 fp
	High		1 tp	1 tp	0.92 tp	0.95 tp
			6.95 fp	7.06 fp	6.28 fp	0.2 fp
3 Outliers	Medium	Uniform	1.88 tp	1.78 tp	0.86 tp	0.45 tp
			6.43 fp	6.29 fp	6.13 fp	0.18 fp
		Clustered	2.2 tp	2.19 tp	0.76 tp	0.17 tp
			6.52 fp	6.58 fp	6.08 fp	0.18 fp
	High	Uniform	2.96 tp	2.94 tp	2.71 tp	1.99 tp
			6.43 fp	6.31 fp	6.11 fp	0.09 fp
		Clustered	3 tp	3 tp	2.69 tp	2.58 tp
			6.5 fp	6.51 fp	6.06 fp	0.06 fp

Table 7.5: Homogeneous Data Distribution, Non Constant Coefficient of Variance

The percentages of true positive and false positive test results are given by:

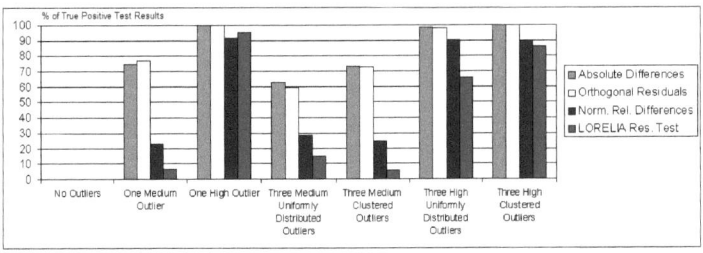

Figure 7.38: Percentages of True Positive Test Results, Non Constant Coefficient of Variance

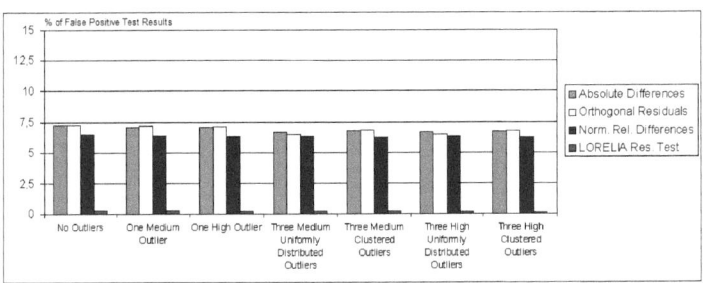

Figure 7.39: Means of True Positive and False Positive Test Results - Homogeneous Data Distribution, Non Constant Coefficient of Variance

The following observations are made:

(i.) The medium outlier term is not well identified.

(ii.) The outlier tests for the absolute differences and for the orthogonal residuals both deliver too many false positive test results which are all located at the end of the measuring range (compare plots in Appendix B) which is explained by the fact that the residual variances are increasing. Although, the percentages of true positive test results are rather high, both test are not appropriate here as the percentages of false positive test results are unacceptable large with about $6.6\% - 7.2\%$ for all data situations.

(iii.) The test based on the normalized relative differences deliver too many false positive test results which are all located at the low concentration range (compare plots in Appendix B) which is explained by the fact that the coefficient of variances is not constant here. Despite the high percentages of true positive test results, both test are not appropriate here as the percentages of false positive test results are unacceptable high with about 6.3% for all data situations.

(iv.) The LORELIA Residual Test delivers the only appropriate results, with respect to the percentages of false positive test results. Again it can be observed that the LORELIA Residual test is more conservative than all other tests. The percentage of true positive test results is maximal for the case of one high outlier term with 95%.

(v.) Again, the percentages of true positive test results are influenced by the position of the outliers which will be discussed in detail in Section 7.2.

7.1.3.3 General Observations and Conclusions

A general performance ranking is given in the following table:

Residual Variance	Absolute Differences	Orth. Residuals	Norm. Rel. Differences	LORELIA Res. Test
Constant	**Highly appropriate,** no adjustment of the local significance levels, thus too sensitive		**Not appropriate,** too many false positive test results at the low concentration range	**Highly appropriate,** Bonferroni adjusted confidence limits
Constant CV	**Not appropriate,** too many false positive test results at the high concentration range		**Highly appropriate,** no adjustment of the local significance levels, thus too sensitive	**Appropriate,** performance depends on the outlier position and on the amount of the increment between the local residual variances, Bonferroni adjusted confidence limits
Non Constant CV	**Not appropriate,** too many false positive test results at the high concentration range		**Not appropriate,** too many false positive test results at the low concentration range	**Appropriate,** performance depends on the outlier position and on the amount of the increment between the local residual variances, Bonferroni adjusted confidence limits

Table 7.6: Performance Ranking

Beside the above performance ranking, a look at the plots (regression plot, scatter plot/residual plot) can reveal supplementary information on the behavior of the different tests, which will be resumed and discussed in the following:

(i.) Generally, the high outlier term is a lot better identified by all tests than the medium outlier term which is often hidden in the main body of the data.

(ii.) The fact if the outliers are uniformly distributed or clustered for a total number of three outliers does not seem to have an observable influence on the test outputs. This is explained by the fact that all test are constructed to be robust against outliers, the LORELIA Residual Test involves a reliability measure in the residual variance estimator, all other tests are based on the median absolute deviation as a robust measure.

(iii.) Note that the position of the outliers within the dataset does have an important influence for the LORELIA Residual Test which can be mainly observed for simulated datasets with three uniformly distributed outliers. The question how the outlier position influences outlier identification for the LORELIA Residual Test will be explored in detail in Section 7.2.

(iv.) In case of a non constant residual variance, the performance of the LORELIA Residual test clearly depends on the fact how well the local residual variances are estimated. For a constant coefficient of variance, the residual variances are underestimated for higher concentrated samples and overestimated at the low concentration range. Thus, outliers in the higher concentration range may therefore not be identified:

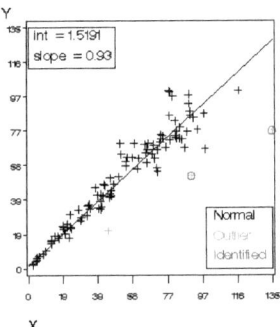 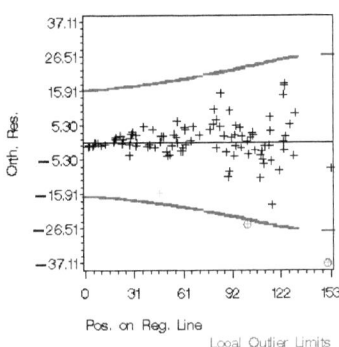

Figure 7.40: Homogeneous Data Distribution, Constant Coefficient of Variance, Three Outliers, Medium Outlier Term, Uniformly Distributed Outliers

(v.) The LORELIA Residual Test is the only test with an adjusted level of significance for the multiple testing situation, which is clearly a drawback of all other tests. Thus, the LORELIA

Residual test is more conservative than all other tests, compare Table 7.2. The residual plots often reveal that a non identified outlier lays just slightly inside its corresponding confidence limit and may be detected when tested with a higher level of significance or a less conservative adjustment procedure than Bonferroni's.

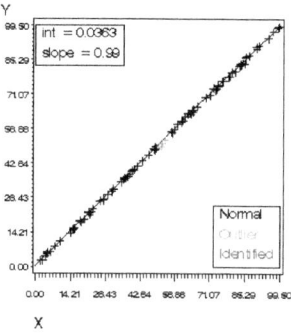
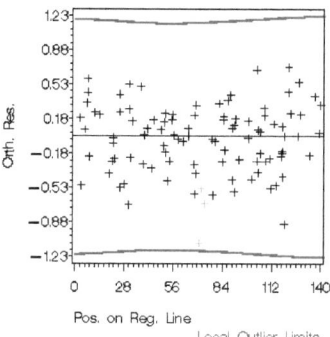

Figure 7.41: Homogeneous Data Distribution, Constant Residual Variance, Three Outliers, Medium Outlier Term, Clustered Outliers

Generally, if the normal assumption for the comparison measure under consideration is met and the sample distribution within the measuring range is homogeneous, the test proposed by [Wadsworth, 1990] delivers appropriate results and is fast and easy to calculate. However these assumptions are often not fulfilled and an appropriate data transformation is not always easy to find. Although there exist a variety of transformation methods proposed in the statistical literature, compare for example [Hawkins, 1980], there still exist data situations for which none of the transformation methods will fit. Moreover in practical applications, the data analysts are often non statisticians who are not very experienced in this field and who often fail to judge the underlying distribution and thus to find an appropriate transformation rule. The LORELIA Residual Test has the clear advantage, that it is globally applicable to any data situation. No statistical assumptions on the underlying error variances have to be checked and no data transformations are needed. The residual plot including the local outlier limits reveals the trend of the underlying residual variance model even to non statisticians.

7.2 Influence of the Outlier Position on its Identification

In this section, the influence of the outlier position within the dataset on its identification is analyzed for different representative data situations by a simulation study. The LORELIA Residual Test is used on simulated datasets containing one outlier. The position of this outlier within the dataset is varied over the entire measuring range. The aim is to generate a functional relation between the outlier position and the expected percentage of true positive test results. The underlying simulation plan will be given in the following section.

7.2.1 Simulation Models

As mentioned above, all simulated datasets will contain one predefined outlier for which the position is varied. The following general data situations will be considered:

Data Distribution	Residual Variance	Outlier Magnitude
Homogeneous	Constant	Medium
		High
	Constant CV	Medium
		High
Inhomogeneous	Constant	Medium
		High
	Constant CV	Medium
		High

Table 7.7: Considered Data Situations to Evaluate the Influence of the Outlier Position

The simulation models will be basically given as in Section 7.1.3.1. Again, every simulated dataset will have a sample size of:

$$n = 100. \tag{7.2.1}$$

The expected measurement values of method M_x and M_y are again assumed to be equal to the true sample concentration:

$$\widetilde{x}_i = \widetilde{y}_i = c_i, \quad \text{for } i = 1, ..., n, \tag{7.2.2}$$

which correspond to the case of equivalent methods M_x and M_y.

The distribution of the true concentrations C_i will be given by:

(i.)

$$C_i \sim U(0, 100), \text{ for } i = 1, ..., n, \tag{7.2.3}$$

to simulate the case of a homogeneous data distribution,

(ii.)
$$C_i \sim logN(0,2), \text{ for } i = 1, ..., n, \quad (7.2.4)$$

to simulate an inhomogeneous data distribution, where the data density decreases with increasing measurement values.

The measurement values including the random errors are thus realizations of:
$$X_i \sim c_i + N(0, \sigma_{r_i}^2) \quad (7.2.5)$$
$$Y_i \sim c_i + N(0, \sigma_{r_i}^2), \quad \text{for } i = 1, ..., n. \quad (7.2.6)$$

Two different residual variance models are considered here:

(i.) The case of a constant residual variance will be modeled by:
$$\sigma_{r_i}^2 \equiv 0.1, \quad \text{for all } i = 1, ..., n. \quad (7.2.7)$$

(ii.) The case of a constant coefficient of variance for the residuals is given by:
$$\sigma_{r_i}^2 = 0.01 \cdot c_i^2, \quad \text{for } i = 1, ..., n. \quad (7.2.8)$$

Outliers will be simulated as described in (7.1.21) to (7.1.27) in Section 7.1.3.1. The number of simulated outliers is fixed to one.

If $x_{(1)}, ..., x_{(100)}$ is the ordered sequence of observed x-values, the outlier term is added to $x_{(j)}$ for a given $j \in \{1, ..., 100\}$. Ideally, a high number of datasets should be simulated for *each* outlier position $(j) \in \{1, ..., 100\}$ and for every data situation under consideration. In order to reduce calculation effort and time, choose a true subset $M \subset \{1, .., 100\}$ of considered outlier positions. The choice of M should depend on the actual distribution C_i of the unbiased measurement values. For homogeneously distributed data, M can be chosen as a homogeneous subset of $\{1, ..., 100\}$. For inhomogeneously distributed data, it is especially important to evaluate the performance of the LORELIA Residual Test in areas with low data density. Therefore, in this case the lowest 10 and the largest 20 outlier positions are additionally included in M:

$$M := \begin{cases} \{(k) : k = 1 + 5 \cdot c, \ c < 20, \ c \in \mathbb{N}\}, & \text{if } C_i \sim U(0, 100), \\ \{(k) : k \leq 10 \text{ or } k > 80 \text{ or } (k = 5 \cdot c, \ c \in \mathbb{N}, \ 3 \leq c \leq 16)\}, & \text{if } C_i \sim logN(0, 2). \end{cases} \quad (7.2.9)$$

Thus, it holds:
$$\#M := \begin{cases} 20, & \text{if } C_i \sim U(0, 100), \\ 44, & \text{if } C_i \sim logN(0, 2). \end{cases} \quad (7.2.10)$$

In conclusion, the simulated datasets are modeled as follows: For all $j \in M \subset \{1,..,100\}$ simulate 500 datasets of the form $(x_1^j, y_1^j), ..., (x_{100}^j, y_{100}^j)$ given as realizations of the following random variables:

$$X_i^j \sim \begin{cases} c_i + N(0, \sigma_{r_i}^2), & \text{if } x_{(i)} \neq x_{(j)}, \\ c_i + N(0, \sigma_{r_i}^2) + out_{x_i}, & \text{if } x_{(i)} = x_{(j)}, \end{cases} \quad (7.2.11)$$

$$Y_i^j \sim c_i + N(0, \sigma_{r_i}^2), \quad \text{for } i = 1, ..., 100. \quad (7.2.12)$$

For each outlier position (j) calculate the percentage of true positive test results with respect to the 500 simulated datasets which contain an outlier at position (j). These percentages are plotted against the outlier position in order to describe a functional relationship. This is done for all data situation presented in Table 7.7.

7.2.2 Homogeneous Data Distribution

To begin with, consider the case of a homogeneous data distribution (7.2.3) for a constant residual variance (7.2.7) or a constant coefficient of variance (7.2.8) with a medium outlier term (7.1.26) or a high outlier term (7.1.27), respectively.

7.2.2.1 Constant Residual Variance

In this section, the most simplest case of a homogeneous data distribution for a constant residual variance is evaluated. To begin with, the expected results will be mathematically estimated and discussed. In the following, these expected results will be compared to the observed simulation results.

Expected Results

In the following, the local confidence limits $C_{\alpha_{\text{loc}}, i}$ will be approximated analytically for the special case of a homogeneous data distribution and a constant residual variance. The following theorem will be used for the approximation:

Theorem 7.4
Consider the case of a constant residual variance:

$$R_i \stackrel{iid}{\sim} N(0, \sigma_r^2), \text{ for } i = 1, ..., n. \quad (7.2.13)$$

Assume further that the following assumption for the reliability measure $\Gamma_{k,n}$ is fulfilled (compare definition (6.5) in Section 6.4):

$$\Gamma_{k,n} = 1, \text{ for all } k = 1, ..., n. \quad (7.2.14)$$

In this case the LORELIA Residual Variance Estimator given in Definition 6.7 in Section 6.5 is an unbiased estimator of the true residual variance σ_r^2.

Proof: The LORELIA Residual Variance estimator is given by:

$$\hat{\sigma}_{r_i}^2 = \frac{1}{\sum_{l=1}^n w_{il}} \cdot \sum_{k=1}^n w_{ik} \cdot r_k^2 = \frac{1}{\sum_{l=1}^n \Delta_{il} \cdot \Gamma_{l,n}} \cdot \sum_{k=1}^n \Delta_{ik} \cdot \Gamma_{k,n} \cdot r_k^2, \quad \text{for } i = 1, ..., n.$$

Note that, by Definition (6.2) in Section 6.4, the distance measure Δ_{ik} is statistically independent of the random variable R_k^2, as the distance between the orthogonal projection does not depend on the actual values of the residuals. The reliability measure $\Gamma_{k,n}$ defined in (6.5) however is statistically dependent of R_k^2. In the following it will therefore be denoted by $\Gamma_{k,n}(R_k^2)$.

It holds:

$$
\begin{aligned}
E\left(\hat{\sigma}_{r_i}^2\right) &= E\left(\frac{1}{\sum_{l=1}^n \Delta_{il} \cdot \underbrace{\Gamma_{l,n}(R_k^2)}_{=1, \text{ by (7.2.14)}}} \cdot \sum_{k=1}^n \Delta_{ik} \cdot \underbrace{\Gamma_{k,n}(R_k^2)}_{=1, \text{ by (7.2.14)}} \cdot R_k^2\right) \\
&= E\left(\frac{1}{\sum_{l=1}^n \Delta_{il}} \cdot \sum_{k=1}^n \Delta_{ik} \cdot R_k^2\right) \\
&= \frac{1}{\sum_{l=1}^n \Delta_{il}} \cdot \sum_{k=1}^n \Delta_{ik} \cdot E\left(R_k^2\right) \\
&= E\left(R^2\right) \cdot \frac{1}{\sum_{l=1}^n \Delta_{il}} \cdot \sum_{k=1}^n \Delta_{ik} \\
&= E(R^2) - \underbrace{E(R)^2}_{=0, \text{ by (7.2.13)}} \quad = Var(R) = \sigma_r^2.
\end{aligned}
$$

Thus, the LORELIA residual variance estimator is unbiased in the case of a constant residual variance σ_r^2 if the reliability measure $\Gamma_{k,n}(R_k^2)$ equals 1. $\square$

Remark 7.5

If the residual variances are assumed to be constant over the measuring range and no outliers are present, then $\gamma_{k,n}$ defined in (6.4.9) in Section 6.4.2 is approximately given by $\frac{1}{n}$ for all $k = 1, ..., n$ and thus $\Gamma_{k,n} \approx 1$ which motivates assumption (7.2.14) in Theorem 7.4.

The local outlier limits are calculated as $(1 - \alpha_{\text{loc}})\%$ approximative confidence intervals $C_{\alpha_{\text{loc}},i}$ for $i = 1, ..., n$. In order to give a rough estimate of the expected percentage of true positive test results, these local confidence intervals will be approximated for every $i = 1, ..., n$ by:

$$
\begin{aligned}
C_{\alpha_{\text{loc}},i} &:= [-t_{DF_i,(1-\frac{\alpha_{\text{loc}}}{2})} \cdot \hat{\sigma}_{r_i}, t_{DF_i,(1-\frac{\alpha_{\text{loc}}}{2})} \cdot \hat{\sigma}_{r_i}] \\
&\approx [-z_{(1-\frac{\alpha_{\text{loc}}}{2})} \cdot \sigma_r, z_{(1-\frac{\alpha_{\text{loc}}}{2})} \cdot \sigma_r].
\end{aligned}
\quad (7.2.15)
$$

where $z_{(1-\frac{\alpha_{loc}}{2})}$ is the $(1-\frac{\alpha_{loc}}{2})\%$ quantile of the standard normal distribution.

Now, the distributional properties of an outlier residual R_i are deduced. Remember, that an outlier at position (j) is simulated as:

$$X_{(j)} \sim c_{(j)} + N(0, \sigma_r^2) + out_{x_{(j)}} = c_{(j)} + N(out_{x_{(j)}}, \sigma_r^2),$$
$$Y_{(j)} \sim c_{(j)} + N(0, \sigma_r^2).$$

For visualization, consider the following plot:

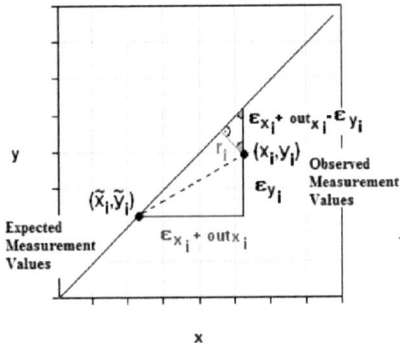

Figure 7.42: The Outlier Residual

By the Pythagorean Theorem the outlying residual is thus a realizations of:

$$\begin{aligned}
R_{(j)} &\sim \frac{1}{\sqrt{2}} \left(N(out_{x_{(j)}}, \sigma_r^2) - N(0, \sigma_r^2) \right) \\
&= \frac{1}{\sqrt{2}} N(out_{x_{(j)}}, 2 \cdot \sigma_r^2) \\
&= N\left(\frac{out_{x_{(j)}}}{\sqrt{2}}, \sigma_r^2 \right) \\
&= N\left(\frac{k \cdot \sigma_r}{\sqrt{2}}, \sigma_r^2 \right) \\
&= \frac{k \cdot \sigma_r}{\sqrt{2}} + N\left(0, \sigma_r^2\right).
\end{aligned} \qquad (7.2.16)$$

Hence, it holds:

$$\begin{aligned}
P\left(R_{(j)} > z_{(1-\frac{\alpha_{loc}}{2})} \cdot \sigma_r\right) &= P\left(\frac{k \cdot \sigma_r}{\sqrt{2}} + N(0, \sigma_r^2) > z_{(1-\frac{\alpha_{loc}}{2})} \cdot \sigma_r \right) \\
&= P\left(N(0, \sigma_r^2) > z_{(1-\frac{\alpha_{loc}}{2})} \cdot \sigma_r - \frac{k \cdot \sigma_r}{\sqrt{2}} \right) \\
&= P\left(N(0, 1) > z_{(1-\frac{\alpha_{loc}}{2})} - \frac{k}{\sqrt{2}} \right).
\end{aligned} \qquad (7.2.17)$$

For $\alpha_{\text{glob}} = 0.1$ and a sample size of $n = 100$, the Bonferroni adjusted local significance level is given by $\alpha_{\text{loc}} = 0.001$. Therefore it follows:

$$P\left(R_{(j)} > z_{(1-\frac{\alpha_{\text{loc}}}{2})} \cdot \sigma_r\right) = P\left(N(0,1) > z_{0.9995} - \frac{k}{\sqrt{2}}\right)$$

$$\approx \begin{cases} P(N(0,1) > 0.46) \approx 0.32, & \text{for } k = 4, \\ P(N(0,1) > -2.37) \approx 0.99, & \text{for } k = 8. \end{cases} \quad (7.2.18)$$

Remark 7.6

Note that the above calculation can only be considered as a rough estimate for the expected results. On the one hand, an outlying residual does not fulfill the assumptions (7.2.13) and (7.2.14) in Theorem 7.4, as the outlying residual follow the distribution given in (7.2.16) and will have a reliability weight which is much smaller than 1. However if only one outlier is present, this problem may be neglected since the outlying measurement is ideally down weighted to an amount of 0 and thus does not bias the residual variance estimator much.

On the other hand, the approximation of the Students-t by the normal quantiles is problematical, as the Students-t quantiles correspond to different degrees of freedom for every $i = 1, ..., n$. By (6.5.5), the degrees of freedom are given by:

$$DF_i = \frac{\left(\sum_{k=1}^{n} w_{ik} \cdot r_k^2\right)^2}{\sum_{k=1}^{n} w_{ik}^2 \cdot r_k^4}, \quad \text{for } i = 1, ..., n,$$

which is an increasing function of the sum of weights $\sum_{k=1}^{n} w_{ik}$. As the weights are based on a continuous distance measure and the data distribution is assumed to be homogeneous, the sum of weights will be maximal for a local residual variance estimate in the middle of the measuring range as the the distances to the neighbored residuals will be the shortest. This will lead to outlier limits which are wider on the borders than in the middle of the measuring range. Thus outliers are expected to be better identified in the middle of the measuring range where the information density is higher than on the borders.

Observed Results

For a medium outlier term which correspond to $out_{r_i} = 4 \cdot \sigma_r$, compare (7.1.26), the plot of the percentages of true positive test results for each outlier position $(j) \in M$ is given by:

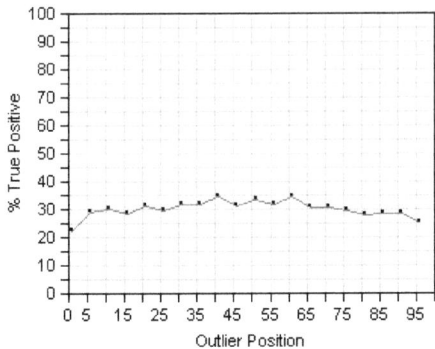

Figure 7.43: Relation between Outlier Position and Percentages of True Positive Test Results - Homogeneous Data Distribution, Constant Residual Variance, Medium Outlier Term

To compare these observed results with the expected results given in (7.2.18) in the previous section, consider the corresponding descriptive analysis:

Quantiles			Moments	
100%	Maximum	34.4	**Mean**	**29.9**
75%	Quartile	31.55	**Standard Deviation**	**2.94**
50%	**Median**	**30.5**	Standard Error for the Mean	0.66
25%	Quartile	28.4	Upper 95% Limit for the Mean	31.27
0%	Minimum	22	Lower 95% Limit for the Mean	28.53
Sample Size				20

Table 7.8: Descriptive Analysis for Percentages of True Positive Test Results - Homogeneous Data Distribution, Constant Residual Variance, Medium Outlier Term

The calculated mean and median for the percentages of true positive test results given by 29.9% and 30.5%, respectively, are close to the expected value of 32% given in (7.2.18). However by Remark 7.6, the above descriptive analysis can only give a rough overview of the simulation results, as the percentages of true positive test results are not expected to be constant over all outlier positions. This also causes the high standard deviation given by 2.94%. If the above plot is zoomed in, it will become obvious that the percentage of true positive test results is higher in the middle of the measuring range than on the borders. To visualize this, a polynomial of degree 2 is fit in the plot:

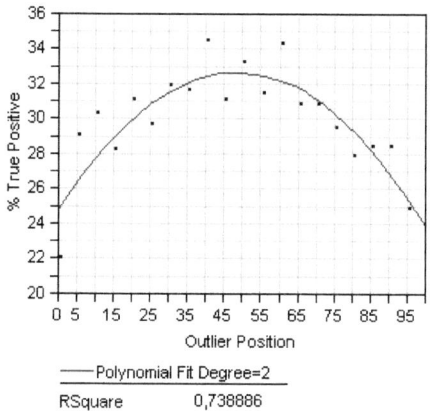

Figure 7.44: Polynomial Fit for Percentages of True Positive Test Results - Homogeneous Data Distribution, Constant Residual Variance, Medium Outlier Term

Now, the functional relation between the outlier position an the percentages of true positive test results is shown for the case of a high outlier term given by $out_{x_i} = 8 \cdot \sigma_r$ which corresponds to (7.1.27).

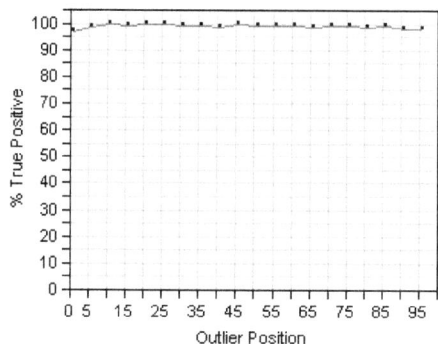

Figure 7.45: Relation between Outlier Position and Percentages of True Positive Test Results - Homogeneous Data Distribution, Constant Residual Variance, High Outlier Term

For the comparison to the expected results consider the corresponding descriptive analysis:

	Quantiles		Moments	
100%	Maximum	99.8	**Mean**	**98.92**
75%	Quartile	99.4	**Standard Deviation**	**0.74**
50%	**Median**	**99.1**	Standard Error for the Mean	0.17
25%	Quartile	98.65	Upper 95% Limit for the Mean	99.27
0%	Minimum	96.6	Lower 95% Limit for the Mean	98.57
Sample Size				20

Table 7.9: Descriptive Analysis for Percentages of True Positive Test Results - Homogeneous Data Distribution, Constant Residual Variance, High Outlier Term

The mean and the median given by 98.92% respective 99.1% are very similar to the expected result of 99% given in (7.2.18). The standard deviation is very low here with 0.74%. This is explained by the fact that a high outlier term corresponds to a very extreme outlier which will nearly always be detected. As the percentage of true positive test results is almost 100% for all outlier positions, the influence of the different local degrees of freedom can thus be neglected here.

In conclusion, the following general observations can be made:

(i.) In the case of a homogeneous data distribution and a constant residual variance, outliers are well identified by the LORELIA Residual if the outlier is well separated from the main body of the data.

(ii.) For an outlier term between $4 \cdot \sigma_r$ and $8 \cdot \sigma_r$, the percentages of true positive test results lay approximatively within $[30\%, 100\%]$.

(iii.) The variation between the percentages of true positive test results for different outlier positions decreases for an increasing outlier term.

7.2.2.2 Constant Coefficient of Variance

Now, the case of a constant coefficient of variance will be considered which is simulated as described in (7.2.8). As the behavior of the LORELIA Residual Variance estimator is much more complex in this case, the expected results can not be explicitly calculated anymore. However in the following section, the expected trends will be discussed based on theoretical considerations.

Expected Results

The LORELIA Weighting Method does not involve any model information of the underlying residual variances in order to be globally applicable for every data situation. Therefore, in the case of a non constant residual variance, the local variance estimates will be smoothed as already mentioned in Section 7.1.3.2. The local LORELIA Outlier Limits will well represent the trend of the underlying residual variance model, but the corresponding variance estimates will be biased. In the case of a constant coefficient of variance, the local residual variance will be overestimated at the low concentration range and underestimated for higher concentrations. Therefore, outliers at the low concentration range are less easily detected than outliers corresponding to higher concentrations. Thus, a monotonously increasing functional relationship between the outlier position and the percentages of true positive test results is expected.

Observed Results

To begin with, consider the case of a medium outlier term. As expected, the percentage of true positive test results is an increasing function of the outlier position. At a low concentration level, the medium outlier term is never identified, whereas for the highest concentration the percentage of true positive test results reaches nearly 70%:

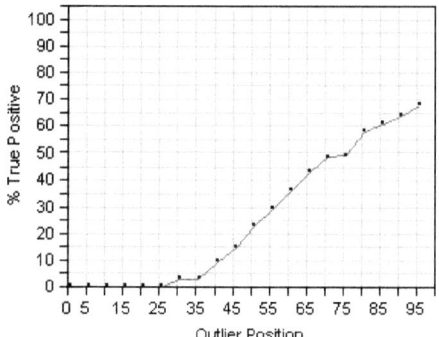

Figure 7.46: Relation between Outlier Position and Percentages of True Positive Test Results - Homogeneous Data Distribution, Constant Coefficient of Variance, Medium Outlier Term

For a high outlier term, the same expected trend can be observed. However, the outlier identification fails completely only for very low concentrations. The percentage of true positive test results increases very soon up to 100%.

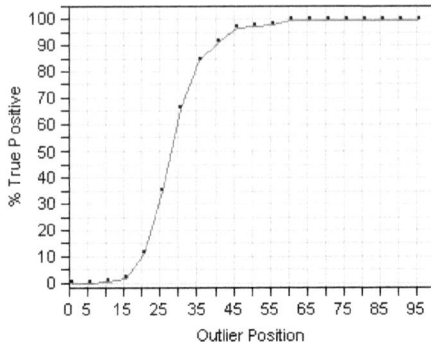

Figure 7.47: Relation between Outlier Position and Percentages of True Positive Test Results - Homogeneous Data Distribution, Constant Coefficient of Variance, High Outlier Term

For the case of a homogeneous data distribution with a constant coefficient of variance, the following general conclusions can be drawn:

(i.) The percentage of true positive test results is an increasing function of the outlier position. The magnitude of this increment depends on the underlying amount of change between the local residual variances. The function becomes steeper if the outlier term increases.

(ii.) Outliers for low concentrated samples are not (well) identified. For an outlier term given by $4 \cdot \sigma_r$, the LORELIA Residual Test fails completely in nearly $1/3$ of the measuring range. For an outlier term of $8 \cdot \sigma_r$ however, the outlier identification fails only in the lowest $1/10$ of the measuring range.

(iii.) If the outlier term is given by $8 \cdot \sigma_r$, the percentage of true positive test results is about 100% in the upper half of the measuring range.

7.2.3 Inhomogeneous Data Distribution

Now, the case of an inhomogeneous sample distribution (7.2.4) will be considered to evaluate the influence of the local data density on the test performance.

Thereby, note that the outlier position is a relative measure for the outlier location with respect to the ordered sequence of x-values. In the case of a homogeneous sample distribution, the relative position of the outlier corresponds well to the absolute location of the outlier within the measuring range. In the case of an inhomogeneous data distribution however, the outlier position does not match with the absolute location of the outlier within the measuring range. In order to describe the functional relationship between the absolute location of the outliers and the percentage of true positive test results, an additional plot is needed. In a first step, the measuring range is split into intervals of length 0.5. The x-component of every simulated outlier lays exactly within one of these intervals. Now, count the number of outliers within each interval over all simulated datasets. For every interval, the percentage of identified outliers with respect to the total number of outliers located in this interval is calculated. The percentages of true positive test results can thus be plotted as a step function of the simulated outlier x-components. The range of considered outlier x-components is reduced to $[2.5, 25]$ here, as for the given simulation settings only some isolated outliers had x-components outside this range.

Note, that the percentages of true positive test results within each interval are based on very different sample sizes, as the total number of outliers located in each interval decreases with decreasing data density. Therefore, the scattering of the true positive test results will increase with increasing x-components. To overcome this confounding aspect, a polynomial of degree 6 is fit additionally to clearly visualize the functional trend.

7.2.3.1 Constant Residual Variance

In the case of a constant residual variance, the influence of the outlier position on the test performance will be mainly determined by the effect of inhomogeneous local data densities (compare the results of Section 7.2.2.1). The expected results are discussed in the following section.

Expected Results

For a low local data density, the sum of weights $\sum_{k=1}^{n} w_{ik}$ for a residual r_i located in this data area will be small, since all distance weights Δ_{ik} with $k \neq i$ will be low. By Remark 7.6, the degrees of freedom DF_i are an increasing function of the sum of weights. Thus, outlier limits will be wider in areas with low local data density than in areas with a high data density. Therefore, outliers will be best identified if the local data density is maximal. As the inhomogeneous data distribution is modeled with a log normal distribution, the local data density is low for high concentrations. The data density will be maximal for small concentrations. Note that for very small concentrations near 0, the local data density will decrease, as well.

Observed Results

For a medium outlier term, the expected trend is well met.

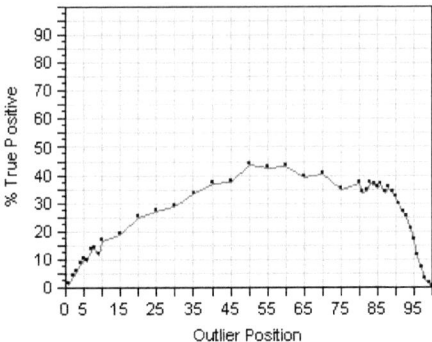

Figure 7.48: Relation between Outlier Position and Percentages of True Positive Test Results - Inhomogeneous Data Distribution, Constant Residual Variance, Medium Outlier Term

The percentages of true positive test results first increase until a maximum is reached. This maximum corresponds to the maximal local data density located at the left border of the measuring range. For increasing concentrations, the local data density decreases and thus the percentage of true positive test results decreases, as well.

Whereas the above plot of the outlier positions only verifies the general trend, the following plot for the absolute location of the outliers clearly shows, that the performance of the LORELIA Residual Test is best for low concentrations, where the local data density is high:

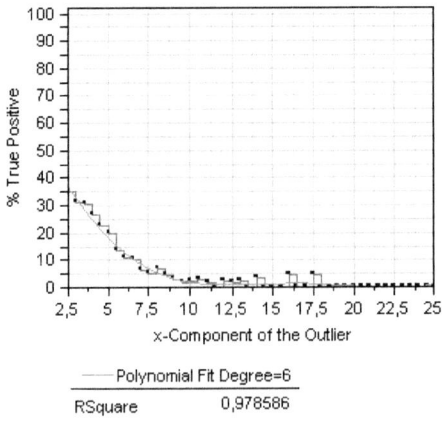

Figure 7.49: Relation between Outlier Location and Percentages of True Positive Test Results - Inhomogeneous Data Distribution, Constant Residual Variance, Medium Outlier Term

For a high outlier term, the general trends remain the same. Outliers within the area of high data density are nearly always identified:

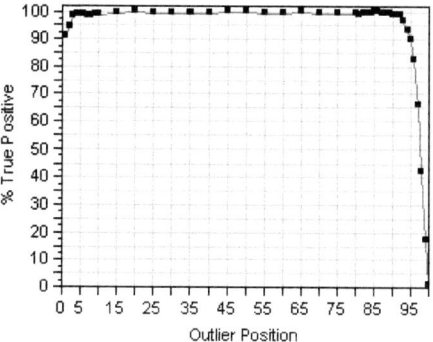

Figure 7.50: Relation between Outlier Position and Percentages of True Positive Test Results - Inhomogeneous Data Distribution, Constant Residual Variance, High Outlier Term

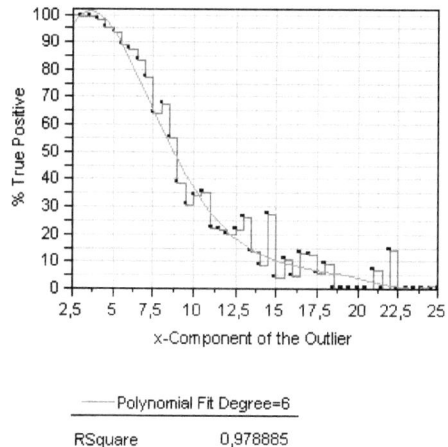

Figure 7.51: Relation between Outlier Location and Percentages of True Positive Test Results - Inhomogeneous Data Distribution, Constant Residual Variance, High Outlier Term

In the case of an inhomogeneous data distribution with a constant residual variance, the following general observations can be made:

(i.) The percentage of true positive test results increases with increasing data density.

(ii.) For a high outlier term given by $8 \cdot \sigma_r$, the percentage of true positive test results is about 100% in the area with the maximal local data density, but decreases down to 0% for the minimal local data density at the right end of the measuring range.

7.2.3.2 Constant Coefficient of Variance

Now, an inhomogeneous sample distribution is considered for the case of a constant coefficient of residual variances.

Expected Results

If the data distribution is inhomogeneous and a constant coefficient of variance is given, the effects discussed in Section 7.2.2.2 and 7.2.3.1 will be mixed. On the one hand, the performance of the outlier test is influenced by the underlying residual variance model, which leads to a monotonously increasing functional relationship between the outlier position and the percentages of true positive test results. On the other hand, the percentages of true positive test results decrease with decreasing data density. As for this simulation model, the residual variances increase with decreasing data density, these effects will be competitive. The observed results will show, which effect has the stronger influence on the local outlier limits.

Observed Results

Consider the case of a medium outlier term. The following plot clearly shows a monotonously increasing functional relationship between the outlier position and the percentages of true positive test results until a certain maximum is reached. In this area, the increasing residual variance model influences the test performance more than the local data density. When the maximum is reached, the local data density overweights the influence of the underlying residual variance model and the percentage of true positive test results decreases (expect for the last outlier position).

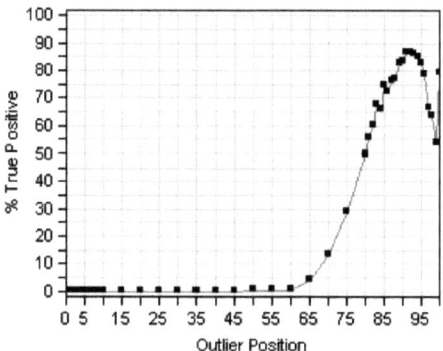

Figure 7.52: Relation between Outlier Position and Percentages of True Positive Test Results - Inhomogeneous Data Distribution, Constant Coefficient of Variance, Medium Outlier Term

The following plot for the absolute location of the outliers shows that the maximum discussed above is located in a low concentration area. Beyond this maximum, the percentages of true positive test results first follow a decreasing trend. At the right end of the considered measuring range, an increasing trend can again be observed. However, the percentages of true positive test results scatter widely for outliers corresponding to larger measuring values as the underlying sample sizes are much smaller. Thus, the increasing trend for high concentrated outliers may be misleading here.

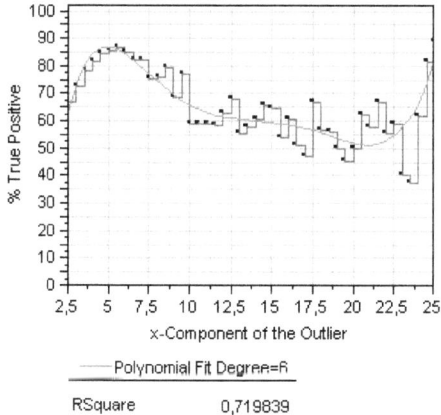

Figure 7.53: Relation between Outlier Location and Percentages of True Positive Test Results - Histogram, Inhomogeneous Data Distribution, Constant Coefficient of Variance, Medium Outlier Term

Similar but more extreme observations as in Figure 7.52 and 7.53 can be made in the case of a high outlier term:

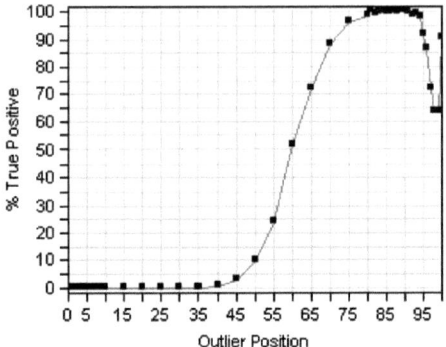

Figure 7.54: Relation between Outlier Position and Percentages of True Positive Test Results - Inhomogeneous Data Distribution, Constant Coefficient of Variance, High Outlier Term

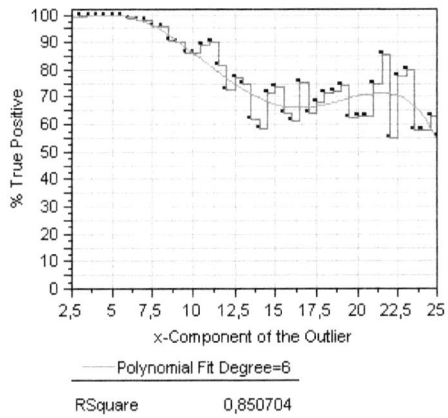

Figure 7.55: Relation between Outlier Location and Percentages of True Positive Test Results - Histogram, Inhomogeneous Data Distribution, Constant Coefficient of Variance, High Outlier Term

Generally, for an inhomogeneous data distribution with a constant coefficient of variance, the functional relationship between the outlier position and the percentage of true positive test results will be a mixture of the functional trends described in Section 7.2.2.2 and 7.2.3.1. The influence balance between the local data density and the underlying residual variance model will depend on the given distributions and parameter settings and thus a functional trend can not generally be described.

7.3 How to Deal with Complex Residual Variance Models

If the local residual variances change too drastically over the measuring range, the performance of the LORELIA Residual Test is very poor. In this section, a discussion on the above problem and a suggestion how to deal with it is given.

If the local residual variances within a dataset are of very different magnitude or the underlying residual variance model is very complex, the LORELIA Residual Variance Estimates may be heavily biased. Therefore, the user of the LORELIA Residual Test is *strongly* recommended to have a look at the corresponding residual plot to verify visually if the local confidence limits seem appropriate and if obvious outlier candidates are well identified. If this is not the case, it may help to split the dataset in order to reduce the complexity of the residual variance model. The following example will illustrate the problem:

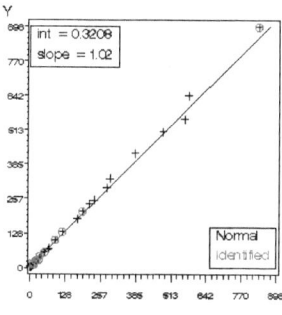 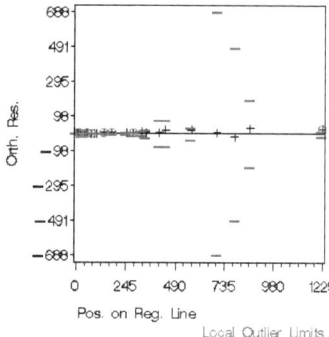

Figure 7.56: Exemplary Dataset - Bad Performance of the LORELIA Residual Test

A cloud of outliers is identified at the low concentration limit. The local outlier limits do not merge smoothly. The reliability plot shows, that a huge number of residuals is down weighted to an amount of 0. These residuals are thus excluded from the calculation of all residual variance estimates:

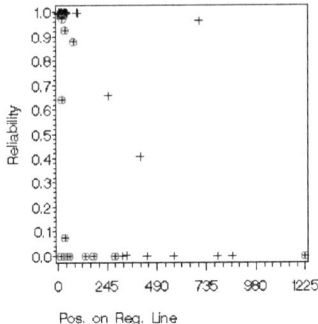

Figure 7.57: Exemplary Dataset - Reliability Plot

The above effect occurs for the following reasons: The sample size is very large with $n = 774$. The sample distribution within the measuring range is very inhomogeneous - about 94% of the measurement values lay within the first hundredth of the measuring range. Moreover, the residual variances are increasing over the measuring range. Thus, the residuals corresponding to the few high concentrated measurement values nearly all are assigned to a reliability weight of 0.

Now, the dataset is split. For the low part of the dataset corresponding to the first 726 measurement values, the local residual variances turn out to be nearly constant. One obvious outlier is clearly identified:

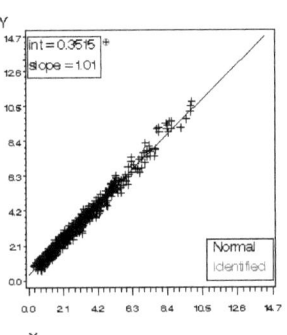 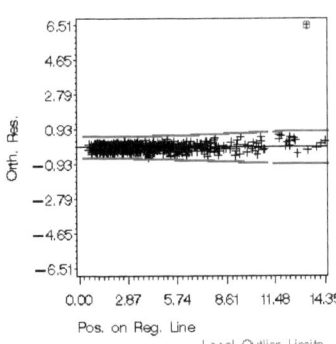

Figure 7.58: Exemplary Dataset, Low Part - Improved Performance

Most residuals correspond to a reliability weight close to 1. Only the identified outlier is down weighted to an amount of 0:

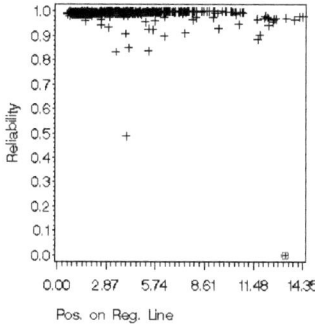

Figure 7.59: Exemplary Dataset, Low Part - Reliability Plot

For the upper part of the dataset, the local residual variances are increasing. No outliers are identified. The local outlier limits merge smoothly:

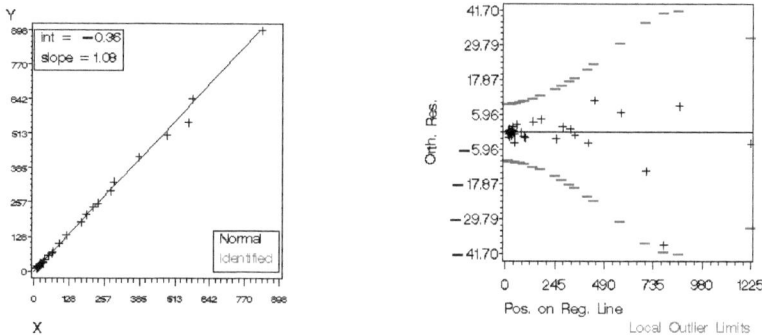

Figure 7.60: Exemplary Dataset, Upper Part - Improved Performance

The reliability plots reveals one value which is down weighted to 0. A look at the residual plot shows that this value lays just slightly within its corresponding confidence interval.

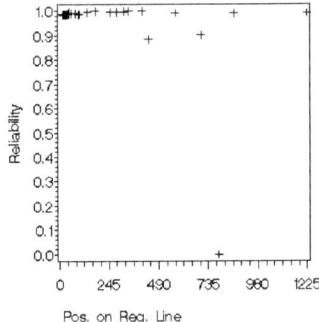

Figure 7.61: Exemplary Dataset, High Part - Reliability Plot

The above example shows, that an appropriate split of the dataset can improve the performance of the LORELIA Residual Test. As the splitting of the dataset implies that two separate outlier tests are performed, the global significance level α_{glob} corresponds no longer to the entire dataset but to the two reduced datasets.

7.4 Considerations on the Alpha Adjustment

Throughout this chapter, the local significance levels of the LORELIA Residual Test were adjusted with the conservative Bonferroni procedure in order to minimize the risk of false positive outlier identifications.

The Bonferroni-Holmes procedure proposed by [Holm, 1979] provides a less conservative adjustment method which can be applied as described in the following:

By (6.2.3) in Section 6.2, the orthogonal residuals approximately follow a Students-t distribution with DF_i degrees of freedom. Thus, each residual can be assigned to a p-values with respect to this distribution. Now, the residuals are ranked in increasing order of the p-values. The outlier test is performed as a stepwise procedure. The first residual $r_{(1)}$ of the ranked sequence is compared to the $(1 - \frac{\alpha_{\text{glob}}}{n})\%$ confidence interval $C_{\frac{\alpha_{\text{glob}}}{n},(1)}$. If $r_{(1)}$ is not identified as an outlier, than none of the the remaining residuals is expected to be an outlier as these residuals correspond to even larger p-values. Thus, the global outlier test can be stopped. If $r_{(1)}$ is identified as an outlier, the second residual $r_{(2)}$ will be compared to the $(1 - \frac{\alpha_{\text{glob}}}{n-1})\%$ confidence interval $C_{\frac{\alpha_{\text{glob}}}{n-1},(2)}$. If $r_{(2)}$ lays within $C_{\frac{\alpha_{\text{glob}}}{n-1},(2)}$, than $r_{(1)}$ corresponds to the only outlier within the dataset and the test will be stopped.

Otherwise $r_{(2)}$ is considered as an outlier, too, and $r_{(3)}$ is compared to $C_{\frac{\alpha_{\text{glob}}}{n-2},(2)}$. This procedure is repeated until $r_{(i)} \notin C_{\frac{\alpha_{\text{glob}}}{n-(i-1)},(i)}$ for an $i = 1, ..., n$ or until all residuals are tested.

In Table 7.2 in Section 7.1.3.2, it has been shown however that in the case of a constant residual variances and a homogeneous sample distribution, the observed actual type 1 error rate of the Bonferroni adjusted LORELIA Residual Test is not conservative but very close to the global significance level for a given sample size of $n = 100$. Therefore, the use of a more complex and less conservative adjustment procedures will not be useful here. However, this may be different for higher sample sizes.

If the underlying residual variance model is not constant, the type 1 error rates do not meet the global significance level as the residual variance estimates are biased due to a smoothing effect. The magnitude of the actual type 1 error rates depend on the amount of this bias. A less conservative adjustment procedure however would be expected to deliver even larger type 1 error rates.

For the above reasons, the choice of the Bonferroni adjustment procedures can be highly recommended in this context. A direct comparison of the actual type 1 error rates between different adjustment methods (for example Bonferroni versus Bonferroni-Holmes) will be the task of future work.

7.5 Summary of the Performance Results

In the previous sections, the performance of the LORELIA Residual Test has been broadly and diversely discussed. The advantages of the new test as well as its limitations were illustrated in examples and by simulation studies. The most important results will be summarized in this section.

Common outlier tests, like the test of [Wadsworth, 1990] presented in Chapter 5 are based on strong statistical assumptions on the comparison measure under consideration. There exist a variety of different transformation formulas in the statistical literature, compare for example [Hawkins, 2002], which allow to apply standard outlier test to the most common data situations in method comparison studies. However, an appropriate data transformation rule is not always easy to find, especially if the data analyst is not very familiar with the different transformation methods. As in many clinical or laboratory applications, the outlier analysis is not performed by statistical experts, this causes serious problems, as an inappropriate data transformation may lead to wrong conclusions about the presence or absence of outliers. Moreover, there exist data situations in which none of the transformation methods proposed in the literature will fit. The new LORELIA Residual Test has the clear advantage to be globally applicable to most data situations in method comparison studies. The data analyst does not need to check special statistical assumptions and no further knowledge on the underlying measurement error model is needed. This provides a clear advantage of the new test, although for some simple data situations, standard outlier tests may be slightly superior.

In the case of a constant residual variance model and a homogeneous sample distribution, the LORELIA Residual Test performs equally good as the standard test proposed by [Wadsworth, 1990] applied on the absolute differences. However, as the test of [Wadsworth, 1990] does not involve an adjustment of the local significance level, an accumulation of type 1 errors occurs, whereas the LORELIA Residual Test is properly adjusted, here by Bonferroni's method.

For non constant residual variances, the local LORELIA Residual Variance Estimates are biased due to a smoothing effect. The magnitude of this bias depend on the amount of change between the local residual variances. For most underlying residual variance models, the LORELIA Residual Test still delivers appropriate results. If the residual variances are known to be proportional to the true concentration (constant coefficient of variance), the outlier test of [Wadsworth, 1990] based on the normalized relative differences outperforms the LORELIA Residual Test slightly. This can be regarded as the price for the model independent approach of the new test.

As the LORELIA Residual Test is a local outlier test, the identification of an outlier depends on its position within the measuring range:
On the one hand, the identification of outliers is influenced by the smoothing effect, which occurs for non constant residual variance models. For increasing residual variances, the LORELIA Residual Variances are underestimated at the low concentration range and overestimated for higher concentrations. In Section 7.2, it is clearly shown that due to this effect, outliers are much better identified for higher concentrated samples. Therefore, outliers in the low concentration range must correspond to a very large outlier term in order to be properly identified, whereas for higher concentrated samples there may occur false positive test results.
On the other hand, the identification of outliers is influenced by the local data density. If the local data density is low, the outlier limits become more conservative and thus existing outliers may not be detected whereas outliers within dense data areas are much easier identified. This however is a desirable effect, as a low data density corresponds to a low level of evidence for the outlier classification. Note, that the local level of data evidence is completely neglected by common outlier, which is a clear drawback. Therefore, in the case of an inhomogeneous sample distribution, the LORELIA Residual Test delivers the more informative results.

As pointed out in Section 7.3, the performance of the LORELIA Residual Test is very poor if the local residual variances change too drastically over the measuring range and if the sample distribution is extremely inhomogeneous. In this case, the local outlier limits do not merge smoothly which can easily be verified visually by having a look at the corresponding residual plot. In this case, a splitting of the dataset may help to reduce the complexity of the underlying residual variance models within the two partial datasets. Applying the LORELIA Residual Test to the new reduced datasets can seriously improve the performance. However, it would be much more satisfying to define formal rules in order to decide *in advance* if a dataset is too inhomogeneous or if the underlying residual variance model is too complex. This will be a task for future work.

Chapter 8
Conclusions and Outlook

Method comparison studies are performed in order to evaluate the relationship between two measurement series, for example to compare two measurement methods, two instruments or two diagnostic tests. Several samples at different concentration levels are measured with both methods or instruments, respectively. Ideally, equivalent methods deliver the same measurement values for each sample. However, both methods are usually exposed to random errors, so the actual measurement values will not exactly be equal.

Method comparison studies are rather evaluated by fitting a linear regression line or by analyzing the measurement differences. Outliers thus correspond to surprisingly large residuals or to measurement values with extremely large differences, respectively. However, what can be interpreted as extreme depends on what is considered as normal, so to say on the underlying distribution of the comparison measure under consideration.

Common outlier tests for method comparison studies like the test proposed by [Wadsworth, 1990] are based on the homoscedastic normal assumption of the respective comparison measure. As the random error variances of the measurement values are often functionally related to the true sample concentration, the homoscedasticity of this normal assumption is often not fulfilled. A variety of different data transformation methods are proposed in the literature, compare for example [Hawkins, 2002], which can be applied in order to obtain a homoscedastic normally distributed comparison measure. However, it is not always easy to find the right data transformation method, especially for non statisticians who are not experienced in the field. Unfortunately, it is common clinical practice that the outlier analysis is not performed by statistical experts. A wrong data transformation however can result in wrong conclusions about the presence or absence of outliers. Moreover, the common transformation methods will only be useful if the random errors in both methods can be described by simple additive or multiplicative models. However, there exist data situations in which none of the transformation methods proposed in the literature will fit.

Another drawback of common approaches is that the local data density is not taken into account.

Datasets in method comparison studies often correspond to a very inhomogeneous sample distribution. Thus, the local level of data evidence to judge a value as an outlier is not equal over the measuring range. Intuitively, existing outliers should be easier identified in areas with a high data density where the local level of evidence is high, whereas for surprisingly extreme observations corresponding to isolated values the local data evidence is low and thus the extremeness of the observation may as well be due to a high local random error variance. If the comparison measure under consideration is assumed to be normally distributed with constant variances over the whole measuring range, the local level of data evidence may be neglected - however the question remains how this assumption can be verified, if the data density is low.

Note that most outlier tests proposed in the statistical literature are constructed to test only a predefined number of outlier candidates. Outlier candidates thereby correspond to the kth most extreme values with respect to the underlying statistical distribution of the population of interest. However, in the case of heteroscedastic error variances and an inhomogeneous sample distribution, it is no longer obvious which values correspond to the most extreme observations as the underlying model of the error variances is unknown. Therefore, outlier tests which test only a predefined number of outlier candidates can not be applied in this context.

As method comparison studies are often evaluated by fitting a linear regression line, it may seem appropriate to consider the various tests proposed in the literature to identify outliers from the linear model, compare for example [Rousseeuw, Leroy, 1987]. However, these test search for values with a high influence on non robust parameter estimates rather than for true outliers which correspond to extremely large residuals with respect to a robustly estimated regression line. These so called 'leverage points' have been discussed in Section 3.3.3. As datasets in method comparison studies often show an inhomogeneous sample distribution, such isolated leverage points are commonly met. However, it is not obvious to decide if an isolated leverage point truly is an outlier, as the local level of data evidence is low. Therefore, standard tests which identify leverage points in a linear model are not appropriate in this context.

So far there exist no satisfying solution to the problem of outlier classification in method comparison studies for the case of heteroscedastic random error variances and an inhomogeneous sample distribution.

In this work, a new outlier identification test for method comparison studies based on robust regression was proposed to overcome the special problem of heteroscedastic residual variances and to include the information of the local data density. The new LORELIA Residual Test (=LOcal RELIAbIlity) is based on a local, robust residual variance estimator, given as a weighted sum of the observed residuals. Outlier limits are estimated from the actual data situation without making assumptions on the underlying residual variance model.

In Chapter 7, the performance of the LORELIA Residual Test was evaluated. The new test was compared to common outlier tests for method comparison studies proposed in the literature. Thereby, the outlier test proposed by [Wadsworth, 1990], which was presented in Chapter 5, was chosen as the reference procedure in this work, as it is one of the few outlier tests which scans the whole dataset for the presence of outliers. Therefore its results can directly be compared to the results of the new LORELIA Residual Test. However, the limitations and problems which can be met for the test of [Wadsworth, 1990] are rather general and will be similar for other outlier tests proposed in the literature.

The test comparison showed, that the LORELIA Residual Test is clearly applicable to a much wider range of data situations. No special statistical assumptions have to be checked and no further knowledge on the underlying measurement error model is needed. Therefore the new test is much simpler in use than standard tests which require different data transformations for each random error model. In Section 7.1.1, it has been shown in examples that the LORELIA Residual Test identifies visually suspicious values truly as outliers, independently of the underlying data situations.

In Section 7.1.2 in Theorem 7.2, the superiority of the new test has been theoretically proven for datasets belonging to a simple data model class M.

A simulation study in section 7.1.3 showed that the new test is highly appropriate for the most common error variance models. For some simple variance models, the test of [Wadsworth, 1990] is slightly superior to the new test, however this can be regarded as the price for the new model independent approach. As the test of [Wadsworth, 1990] does not involve an adjustment of the local significance level, its actual type 1 error becomes huge for high sample sizes, whereas the LORELIA Residual Test is much more conservative as it is adjusted by Bonferroni's method.

If the sample distribution is inhomogeneous, the LORELIA Residual Test reacts to the local data density whereas the test of [Wadsworth, 1990] ignores the local level of data evidence and should thus be interpreted with care. In Section 7.2, it was shown that existing outliers are identified best in areas with maximal data density. Moreover, it was demonstrated that the local LORELIA Residual Variance Estimates are biased due to a smoothing effect if the underlying residual variance model is not constant.

The LORELIA Residual Test is applicable to most method comparison datasets which can be met in the clinical context. However, its performance is not equally good for all data situations. The performance is mainly influenced by:

(i.) The sample size,

(ii.) The magnitude of existing outliers,

(iii.) The complexity of the underlying error variance model,

(iv.) The underlying sample distribution.

The problem of the sample size is due to the use of the Bonferroni correction in the multiple test situation which will make the new test very conservative for high sample sizes. This may be overcome be the use of a more complex adjustment of the local significance levels, for example by the procedure of Bonferroni-Holmes discussed in Section 7.4.

It is obvious, that existing outliers can only be identified if they are well separated from the main body of the data. However, the problem of the outlier magnitude is also related to the smoothing effect discussed above. An upwards bias in the estimate of the local residual variance can avoid outlier identification, even if the outlier candidate is well separated.

A large smoothing effect occurs if the local residual variances differ extremely over the measuring range and if the sample distribution within the measuring range is very inhomogeneous. In this case, it can help to split the dataset and to use the LORELIA residual test on the new reduced datasets. This approach was presented in Section 7.5.

Although, the performance of the new LORELIA Residual Test is limited by several criteria, the examples and simulations given in this work demonstrate its wide range of application. It will be the task of future work to formulate explicit conditions, which have to be met in order to guarantee a certain level of performance. These conditions may be given as:

(i.) A limiting maximal value for the allowed sample size,

(ii.) A limiting minimal value for the size of an identifiable outlier,

(iii.) An explicit measure for the complexity of the underlying residual variance model and for the amount of change between the local residual variances,

(iv.) An explicit measure for the inhomogeneity of the sample distribution.

Sample size limitations should be discussed in the context of different adjustment methods for the local significance levels. A measure for the complexity of the underlying residual variance model may be based on the ratio between the largest and the smallest observed residual. A measure for the inhomogeneity of the sample distribution could be given as the percentage of observations laying within a predefined small area, for example within the lowest tenth of the measuring range.

Beside the formulation of explicit conditions for the LORELIA Residual Test, there exist other interesting questions in the field of method comparison studies which may be solved with a similar approach. For example, it would be interesting to formulate an outlier test if more than two methods are to be compared simultaneously. Another interesting task would be to expand the LORELIA Residual Test to datasets which are described by non linear models.

This work provides a widely applicable solution to the problem of outlier identification in method comparison studies. However, there exist various possibilities to expand and to improve this new approach. The field of outlier detection in method comparison studies still offers many interesting problems and questions for further research.

Appendix A

Software Development and Documentation

All program code developed in the context of this work was implemented by the author in SAS® 9.1. The resulting SAS®-programs are saved on the attached disk [1]. A *html-* documentation was produced with the open source documentation software Doxygen 1.5.8, compare [Doxygen, van Heesch, 2008], in order to simplify the program overview and description. The *html-*documentation can be opened over the following path on the attached disk:

...\Program_Documentation\doc\index.html .

The documented programs include:

(i.) An implementation of the LORELIA Residual Test,

(ii.) Implementations of the global outlier tests presented in Chapter 5,

(iii.) Implementations of the simulation studies described in Chapter 7, Sections 7.2 and 7.1.3.

Note that the original source code of the LORELIA Residual Test includes company internal procedures to calculate the Passing-Bablok regression estimators. These procedures can not be published here to preserve the property rights. Therefore some programs can not be run without further implementations, which is explicitly indicated in the respective program description within the documentation. However, an alternative implementation of the LORELIA Residual Test is provided, for which the regression estimators are handled as input parameters. This program can therefore directly be run.

The following list shows all documented files with brief descriptions. More detailed informations on the different programs and their hierarchical structure are given in the *html-*Doxygen-documentation on the attached disk:

---

[1] The program code and its documentation can be achieved from the author on request: *geraldine_r@web.de*

Calculate_OrthRes(SIM).sas [code]	Calculation of the orthogonal residuals for Passing-Bablok regression (version for simulated datasets containing predefined outliers)
Calculate_OrthRes.sas [code]	Calculation of the orthogonal residuals for Passing-Bablok regression
Calculate_OrthRes_withoutRegression.sas [code]	Calculation of the orthogonal residuals for Passing-Bablok regression
doxygroups.sas [code]	Initialization of the Program Modules
Export.sas [code]	Export of the Results
GlobalOutlierTest(SIM).sas [code]	Global outlier tests based on the absolute differences, on the normalized relative differences and on the orthogonal residuals for simulated datasets containing predefined outliers - Note: This program uses ROCHE-internal source code for the Passing-Bablok-Regression and can not be run without further implementations
GlobalOutlierTests.sas [code]	Global outlier tests based on the absolute differences, on the normalized relative differences and on the orthogonal residuals - Note: This program uses ROCHE-internal source code for the Passing-Bablok-Regression and can not be run without further implementations
Import.sas [code]	Data import
LORELIA(SIM).sas [code]	Main program of the LORELIA Residual Test for simulated datasets containing predefined outliers Note: This program uses ROCHE-internal source code for the Passing-Bablok-Regression and can not be run without further implementations
LORELIA.sas [code]	Main program of the LORELIA Residual Test - Note: This program uses ROCHE-internal source code for the Passing-Bablok-Regression and can not be run without further implementations
LORELIA_withoutRegression.sas [code]	The LORELIA Residual Test without the Passing-Bablok Regression Tool- Regression paramters must be given in advance. Note: This program can be run without further implementations
Plot_Data(SIM).sas [code]	Produces regression plot with identified outliers marked in red and simulated outliers marked in orange and saves them to the output directory (version for simulated datasets containing predefined outliers)
Plot_Data.sas [code]	Produces regression plot with identified outliers marked in red and saves them to the output directory
Plot_Data_withoutRegression.sas [code]	Produces regression plot with identified outliers marked in red and saves them to the output directory (version for LORELIA_withoutRegression)
Plot_Residuals(SIM).sas [code]	Produces residual plot with local outlier limits and saves them to the output directory, identified outliers are marked in red, simulated outliers are marked in orange (version for simulated datasets containing predefined outliers)
Plot_Residuals.sas [code]	Produces residual plot with local outlier limits and saves them to the output directory, identified outliers are marked in red
Plot_weight_relia(SIM).sas [code]	Produces reliability plot and saves it to the output directory- for each measurement value, the position of its orthogonal projection on the regression line is assigned to its local reliability, identified outliers are marked in red, simulated outliers are marked in orange (version for simulated datasets containing predefined outliers)
Plot_weight_relia.sas [code]	Produces reliability plot and saves it to the output directory- for each measurement value, the position of its orthogonal projection on the regression line is assigned to its local reliability, identified outliers are marked in red

Simulation_constant.sas [code]	Simulation Studies of Section 7.1.3 'Performance Comparison for Simulated Datasets' and of Section 7.2 'Influence of the Outlier Position on its Identification' - Constant Residual Variance
Simulation_constantCV.sas [code]	Simulation Studies of Section 7.1.3 'Performance Comparison for Simulated Datasets' and of Section 7.2 'Influence of the Outlier Position on its Identification' - Constant Coefficient of Variance
Simulation_nonconstantCV.sas [code]	Simulation Studies of Section 7.1.3 'Performance Comparison for Simulated Datasets' Note: In the simulation study in 7.2 'Influence of the Outlier Position on its Identification' a non constant coefficient of variance is not considered - Non Constant Coefficient of Variance
Weights(SIM).sas [code]	Calculation of the LORELIA Weights (version for simulated datasets containing predefined outliers)
Weights.sas [code]	Calculation of the LORELIA Weights

Figure A.1: Documented SAS® Program Files with Brief Descriptions

Beside the above SAS®-files, several other programs have been developed to guarantee a correct and fast evaluation of the simulations studies. However, as these programs are constructed only to save results and to count events, they do not contain much additional source code. For this reason, these evaluation programs are not documented here.

Appendix B

Test Results of Section 7.1.3

B.1 Constant Residual Variance

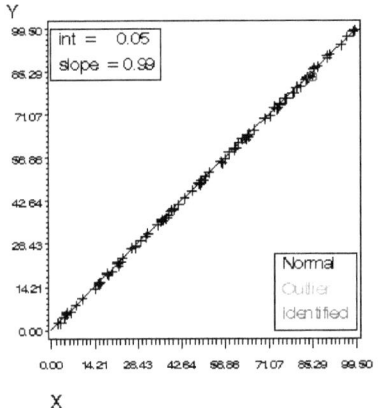
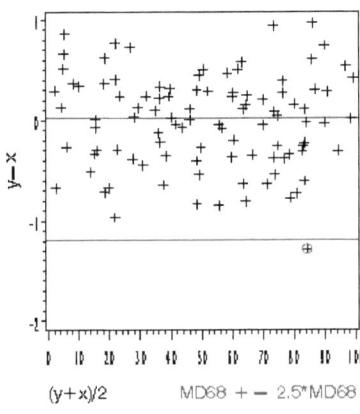

Figure B.1: Simulation 1: Constant Residual Variance, No Outliers - Outlier Test for the Absolute Differences

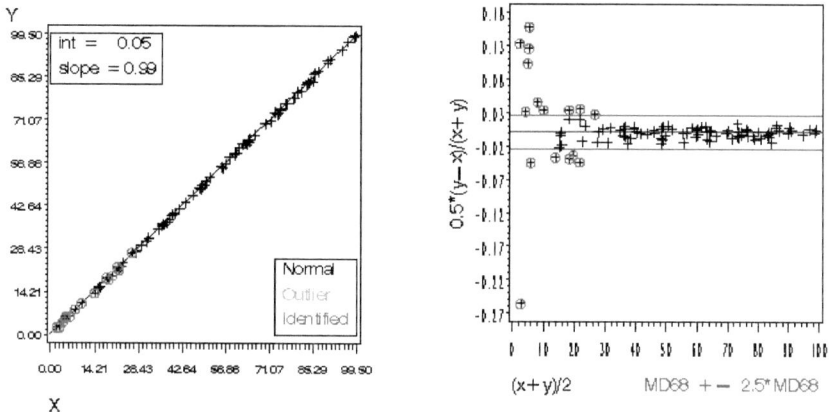

Figure B.2: Simulation 1: Constant Residual Variance, No Outliers - Outlier Test for the Normalized Relative Differences

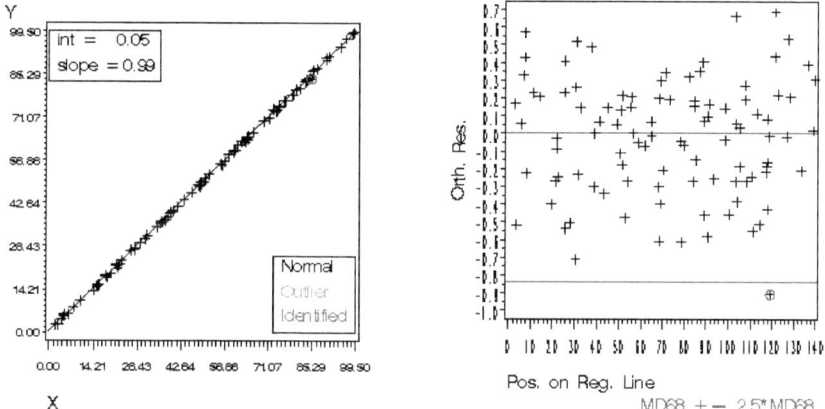

Figure B.3: Simulation 1: Constant Residual Variance, No Outliers - Outlier Test for the Orthogonal Residuals

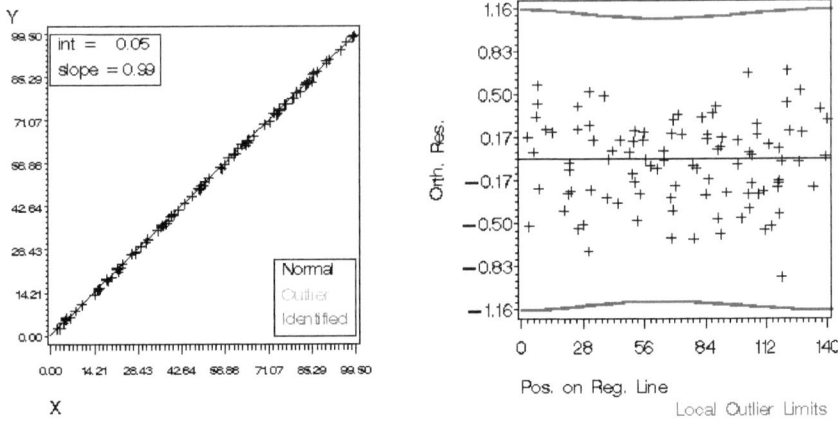

Figure B.4: Simulation 1: Constant Residual Variance, No Outliers -The LORELIA Residual Test

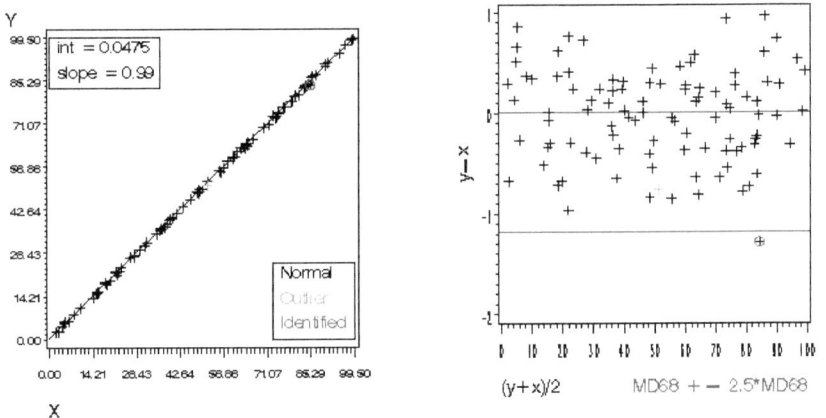

Figure B.5: Simulation 2: Constant Residual Variance, One Outlier, Medium Outlier Term - Outlier Test for the Absolute Differences

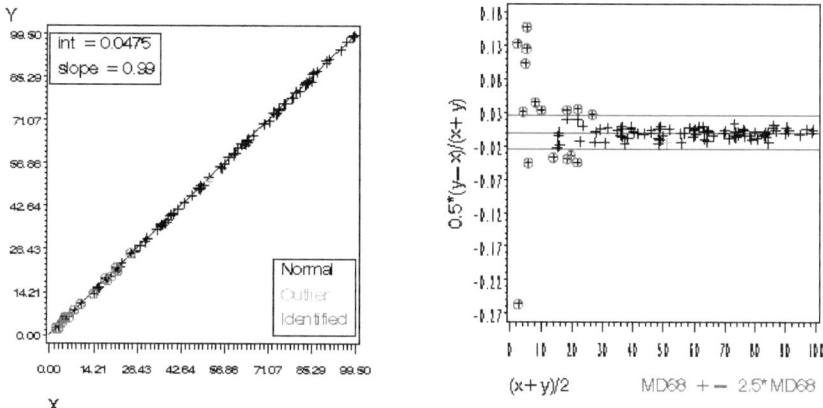

Figure B.6: Simulation 2: Constant Residual Variance, One Outlier, Medium Outlier Term - Outlier Test for the Normalized Relative Differences

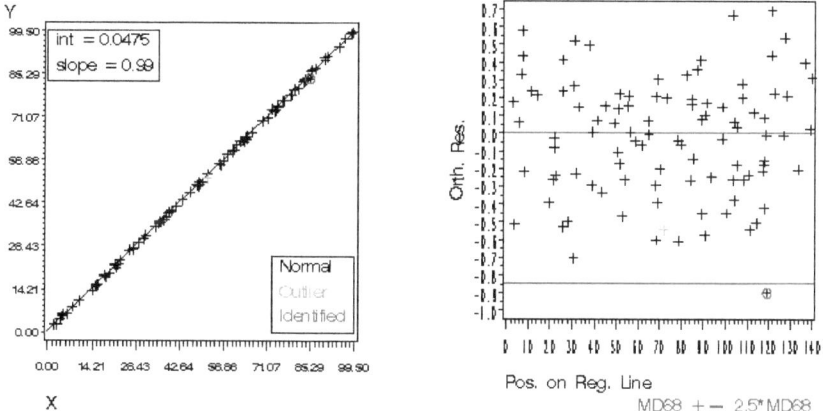

Figure B.7: Simulation 2: Constant Residual Variance, One Outlier, Medium Outlier Term - Outlier Test for the Orthogonal Residuals

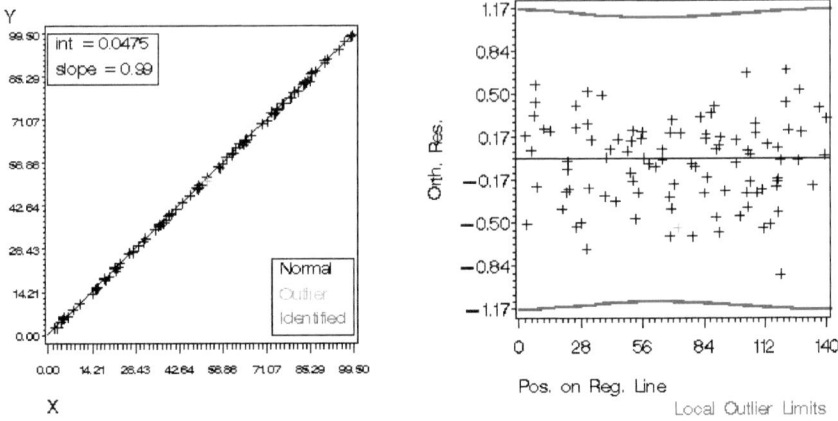

Figure B.8: Simulation 2: Constant Residual Variance, One Outlier, Medium Outlier Term -The LORELIA Residual Test

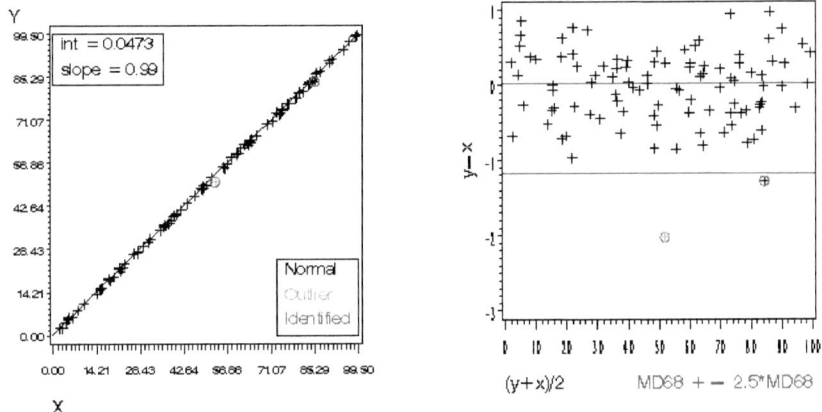

Figure B.9: Simulation 3: Constant Residual Variance, One Outlier, High Outlier Term - Outlier Test for the Absolute Differences

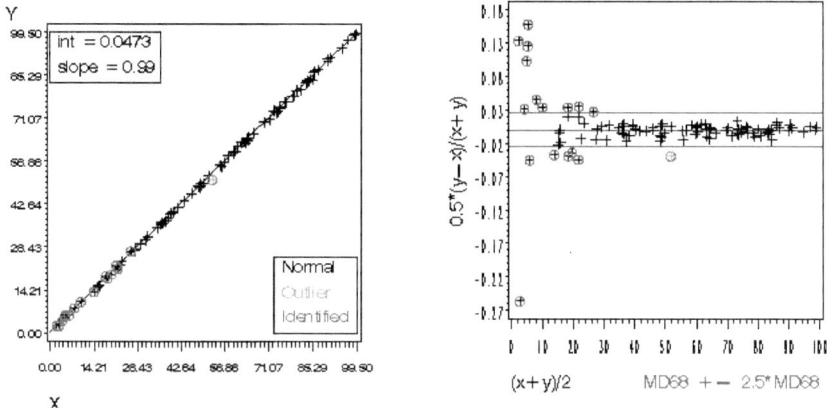

Figure B.10: Simulation 3: Constant Residual Variance, One Outlier, High Outlier Term - Outlier Test for the Normalized Relative Differences

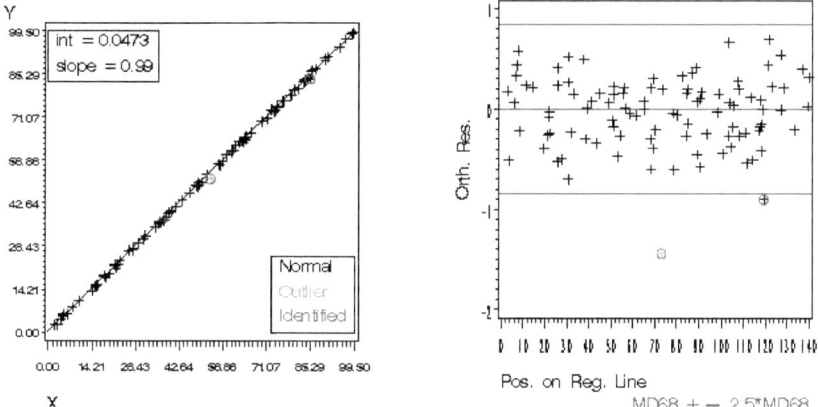

Figure B.11: Simulation 3: Constant Residual Variance, One Outlier, High Outlier Term - Outlier Test for the Orthogonal Residuals

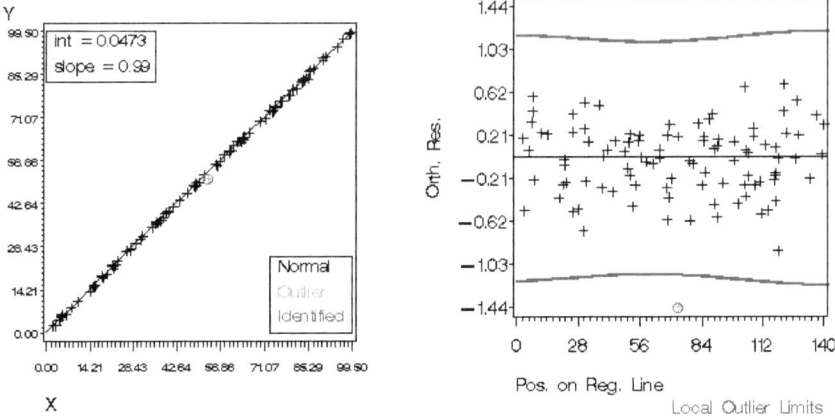

Figure B.12: Simulation 3: Constant Residual Variance, One Outlier, High Outlier Term - The LORELIA Residual Test

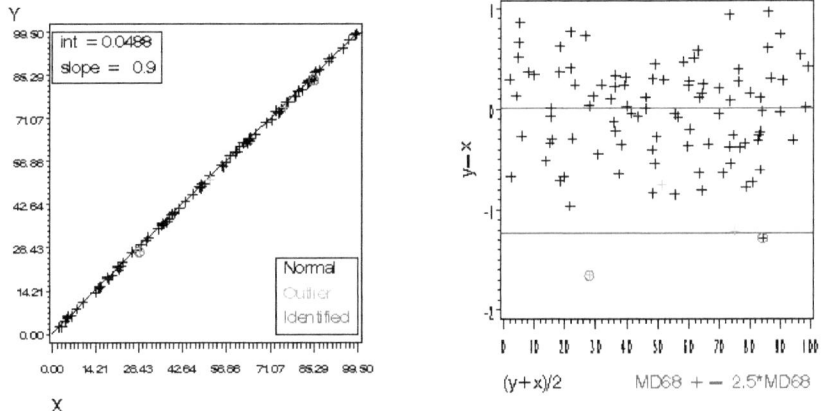

Figure B.13: Simulation 4: Constant Residual Variance, Three Outliers, Medium Outlier Term, Uniformly Distributed Outliers - Outlier Test for the Absolute Differences

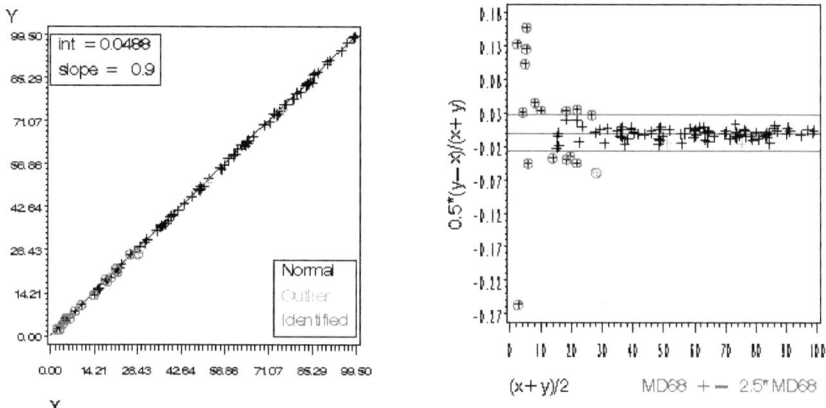

Figure B.14: Simulation 4: Constant Residual Variance, Three Outliers, Medium Outlier Term, Uniformly Distributed Outliers - Outlier Test for the Normalized Relative Differences

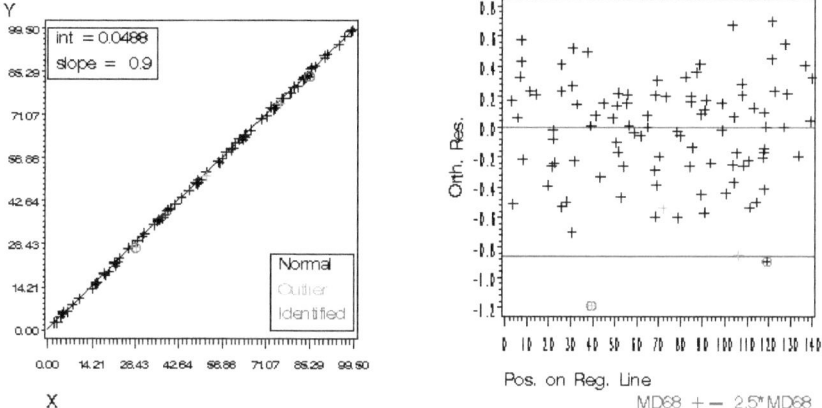

Figure B.15: Simulation 4: Constant Residual Variance, Three Outliers, Medium Outlier Term, Uniformly Distributed Outliers - Outlier Test for the Orthogonal Residuals

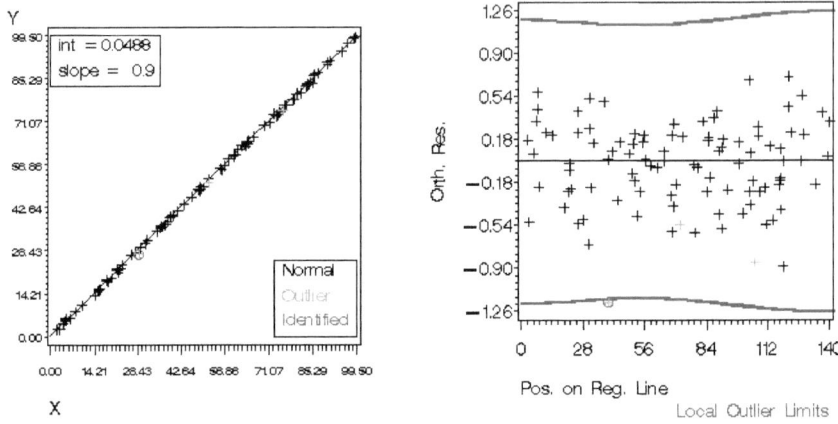

Figure B.16: Simulation 4: Constant Residual Variance, Three Outliers, Medium Outlier Term, Uniformly Distributed Outliers - The LORELIA Residual Test

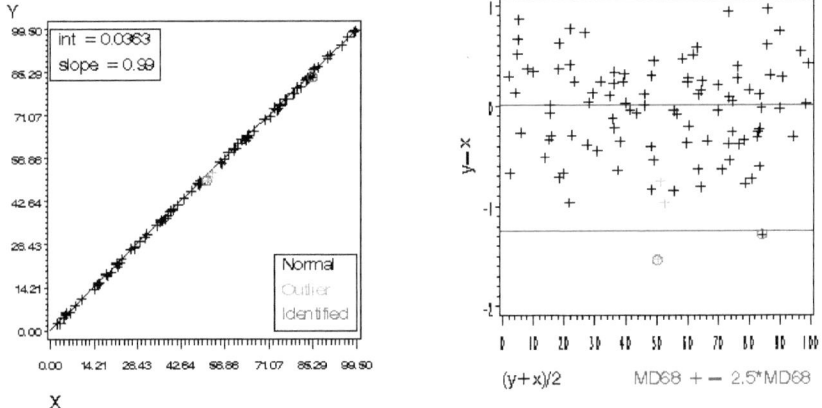

Figure B.17: Simulation 5: Constant Residual Variance, Three Outliers, Medium Outlier Term, Clustered Outliers - Outlier Test for the Absolute Differences

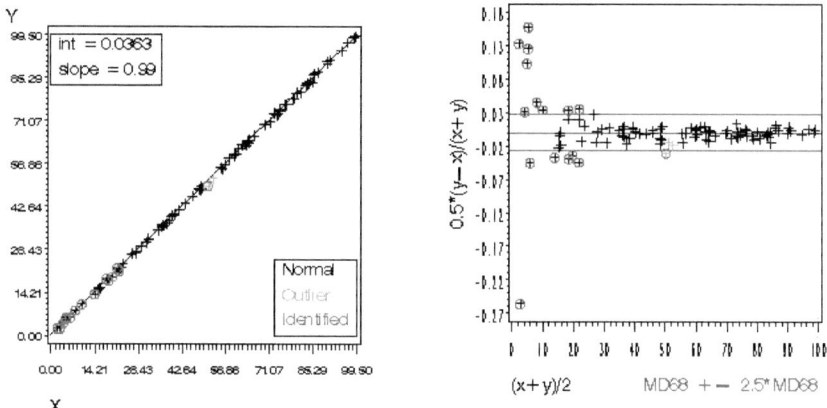

Figure B.18: Simulation 5: Constant Residual Variance, Three Outliers, Medium Outlier Term, Clustered Outliers - Outlier Test for the Normalized Relative Differences

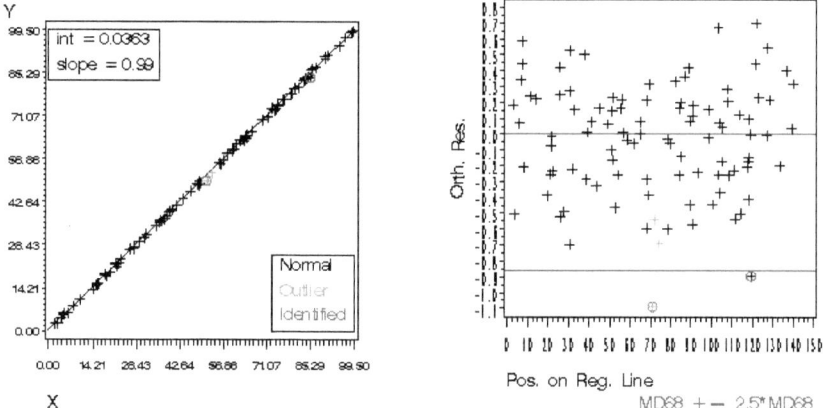

Figure B.19: Simulation 5: Constant Residual Variance, Three Outliers, Medium Outlier Term, Clustered Outliers - Outlier Test for the Orthogonal Residuals

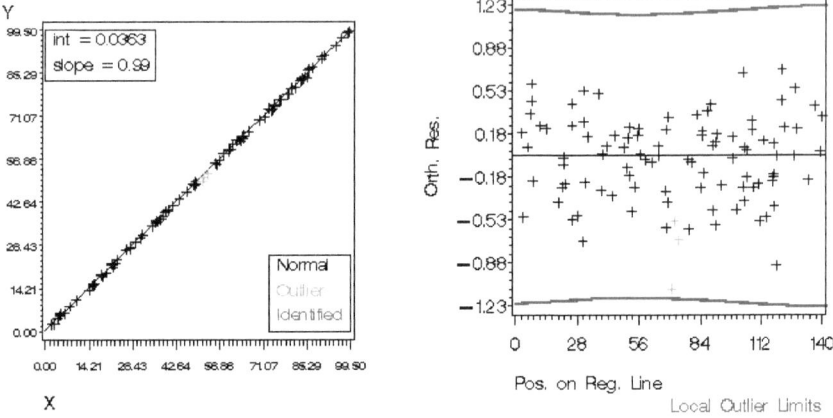

Figure B.20: Simulation 5: Constant Residual Variance, Three Outliers, Medium Outlier Term, Clustered Outliers - The LORELIA Residual Test

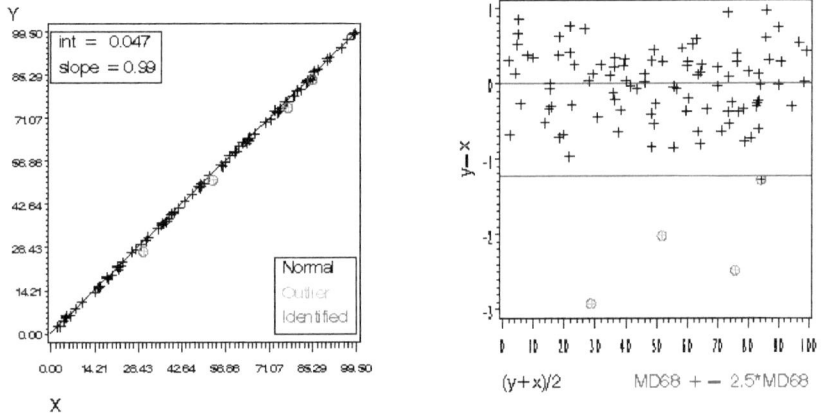

Figure B.21: Simulation 6: Constant Residual Variance, Three Outliers, High Outlier Term, Uniformly Distributed Outliers - Outlier Test for the Absolute Differences

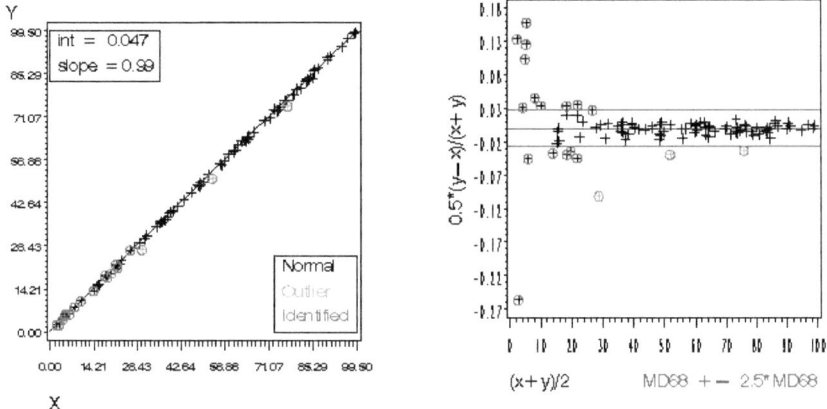

Figure B.22: Simulation 6: Constant Residual Variance, Three Outliers, High Outlier Term, Uniformly Distributed Outliers - Outlier Test for the Normalized Relative Differences

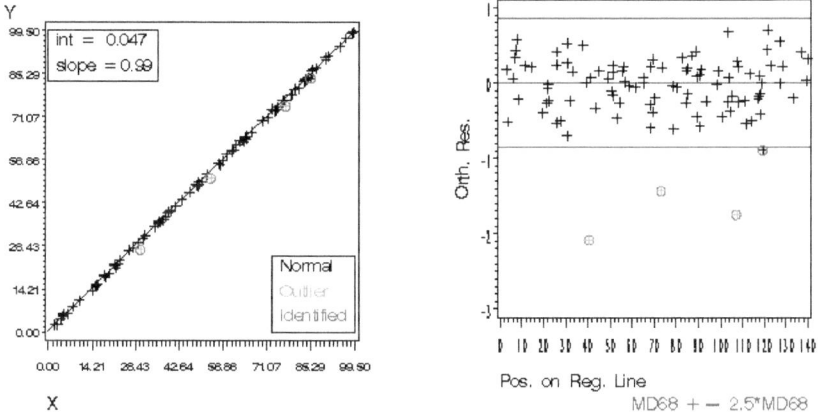

Figure B.23: Simulation 6: Constant Residual Variance, Three Outliers, High Outlier Term, Uniformly Distributed Outliers - Outlier Test for the Orthogonal Residuals

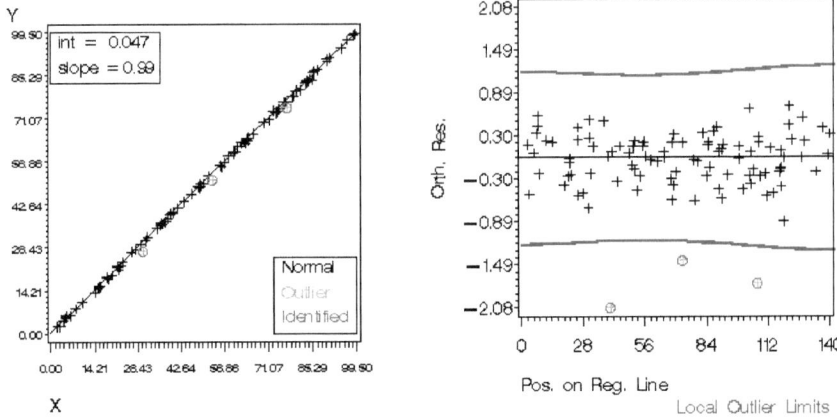

Figure B.24: Simulation 6: Constant Residual Variance, Three Outliers, High Outlier Term, Uniformly Distributed Outliers -The LORELIA Residual Test

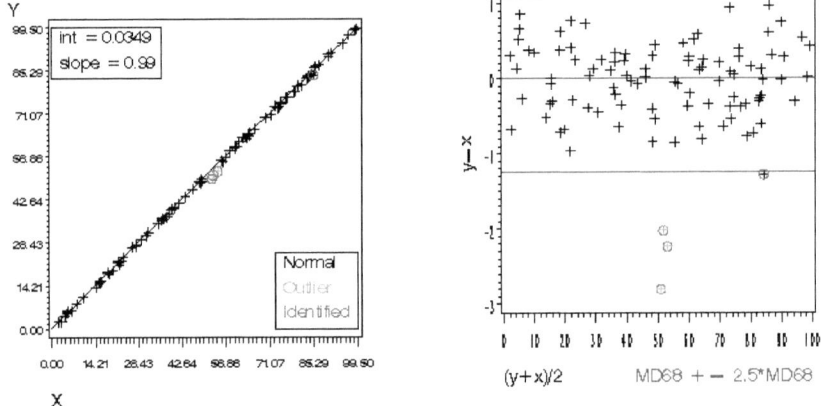

Figure B.25: Simulation 7: Constant Residual Variance, Three Outliers, High Outlier Term, Clustered Outliers - Outlier Test for the Absolute Differences

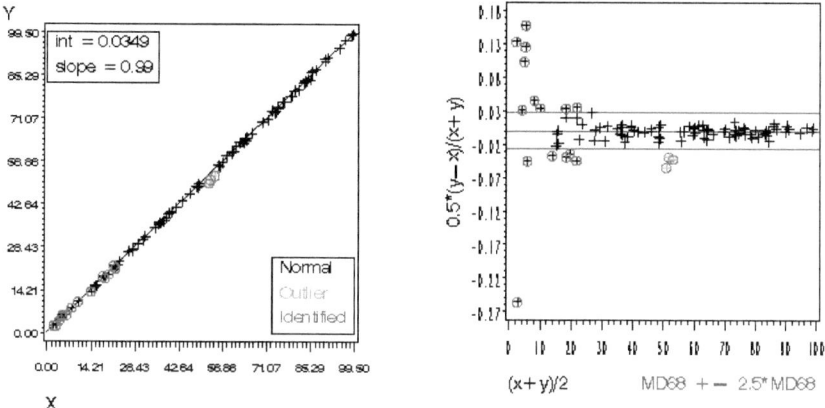

Figure B.26: Simulation 7: Constant Residual Variance, Three Outliers, High Outlier Term, Clustered Outliers - Outlier Test for the Normalized Relative Differences

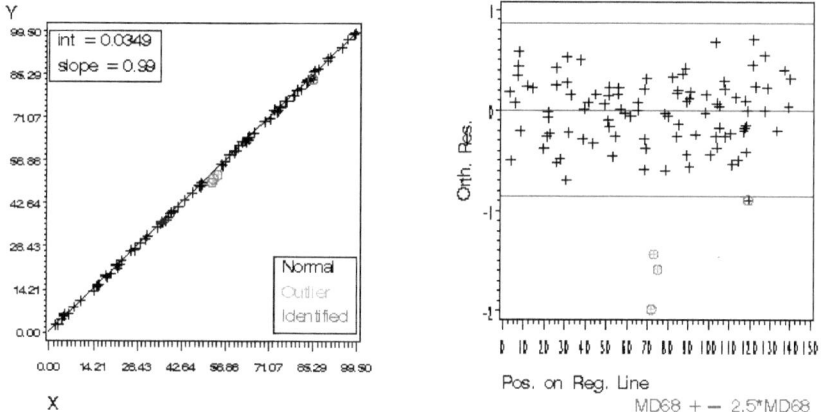

Figure B.27: Simulation 7: Constant Residual Variance, Three Outliers, High Outlier Term, Clustered Outliers - Outlier Test for the Orthogonal Residuals

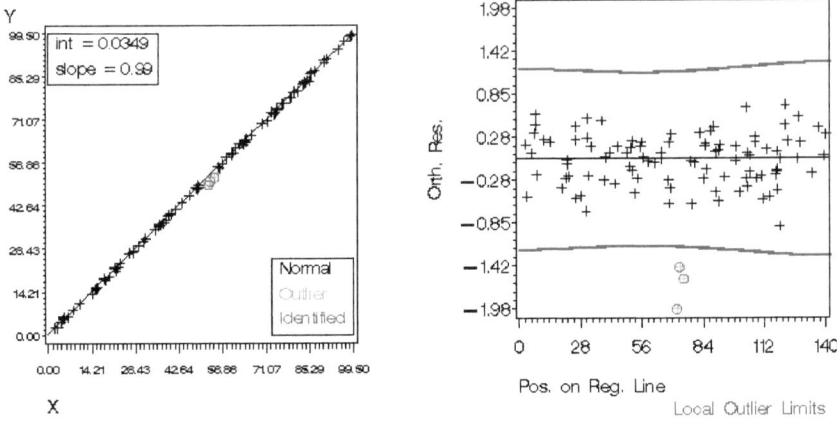

Figure B.28: Simulation 7: Constant Residual Variance, Three Outliers, High Outlier Term, Clustered Outliers -The LORELIA Residual Test

B.2 Constant Coefficient of Variance

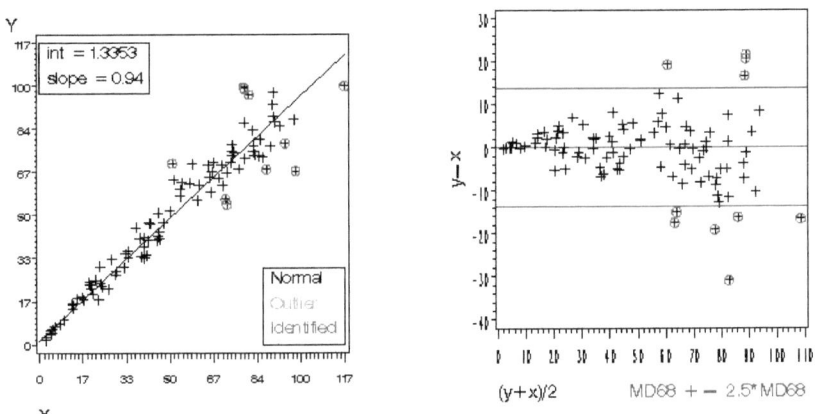

Figure B.29: Simulation 8: Constant Coefficient of Variance, No Outliers - Outlier Test for the Absolute Differences

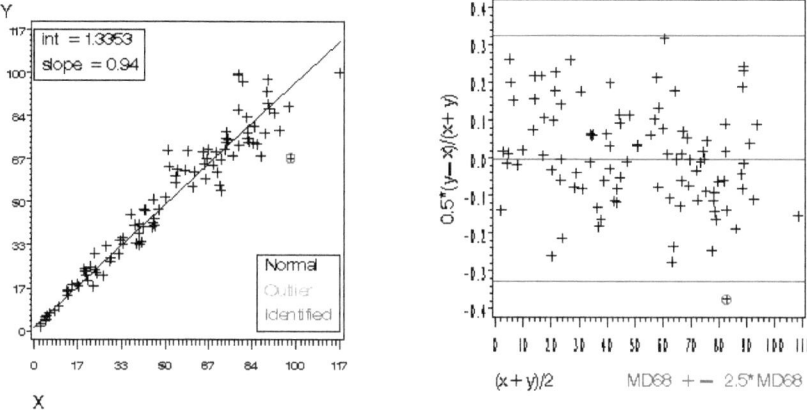

Figure B.30: Simulation 8: Constant Coefficient of Variance, No Outliers - Outlier Test for the Normalized Relative Differences

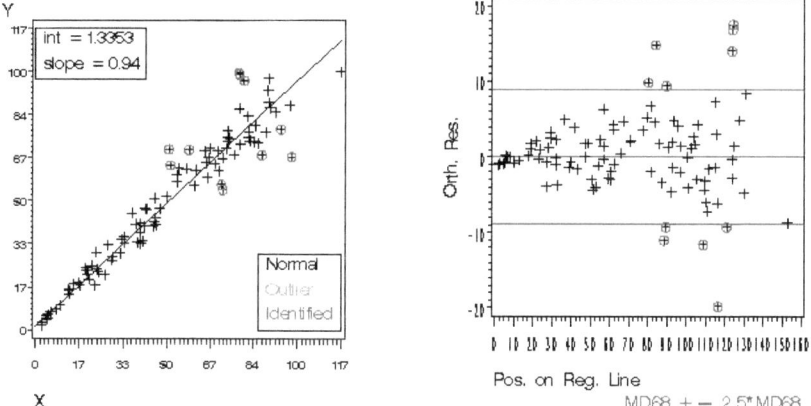

Figure B.31: Simulation 8: Constant Coefficient of Variance, No Outliers - Outlier Test for the Orthogonal Residuals

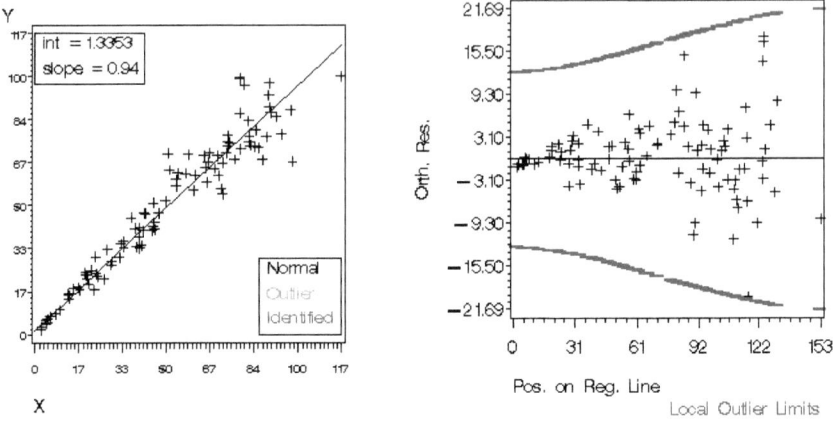

Figure B.32: Simulation 8: Constant Coefficient of Variance, No Outliers - The LORELIA Residual Test

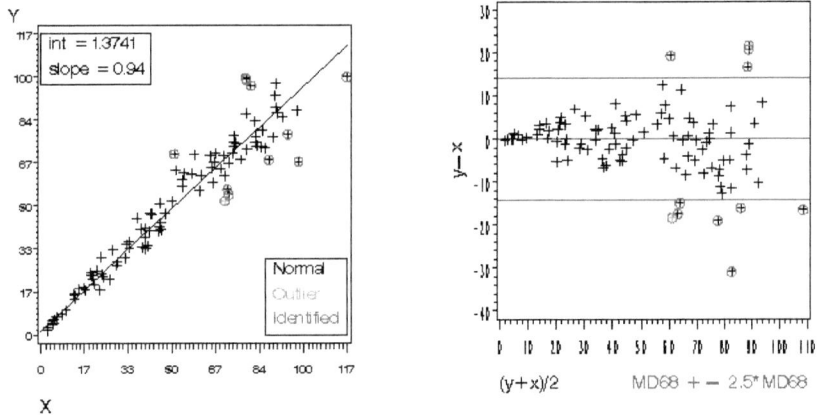

Figure B.33: Simulation 9: Constant Coefficient of Variance, One Outlier, Medium Outlier Term - Outlier Test for the Absolute Differences

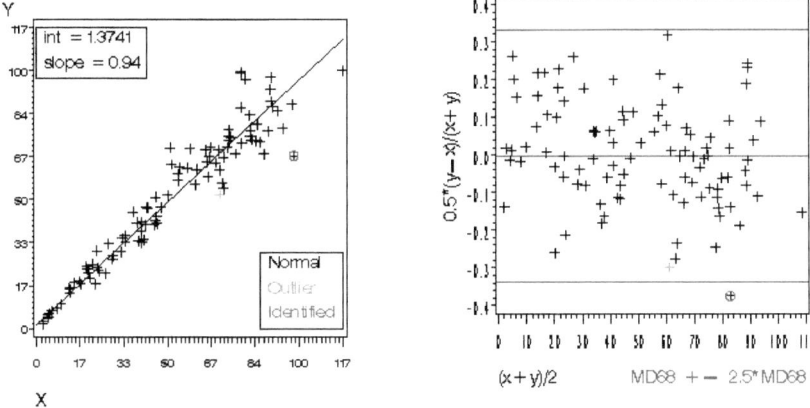

Figure B.34: Simulation 9: Constant Coefficient of Variance, One Outlier, Medium Outlier Term - Outlier Test for the Normalized Relative Differences

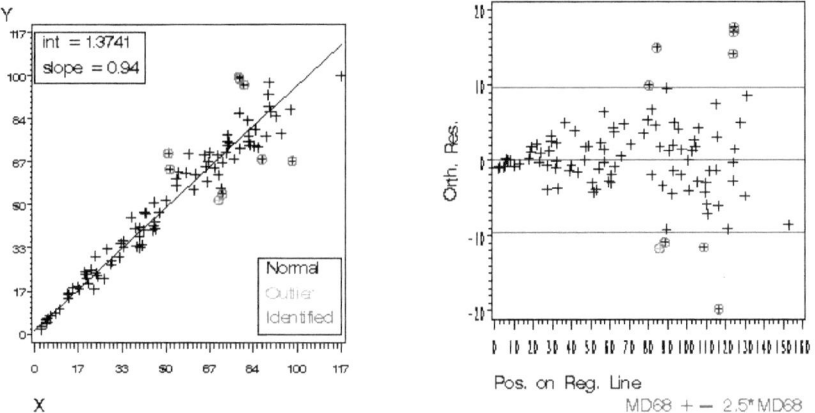

Figure B.35: Simulation 9: Constant Coefficient of Variance, One Outlier, Medium Outlier Term - Outlier Test for the Orthogonal Residuals

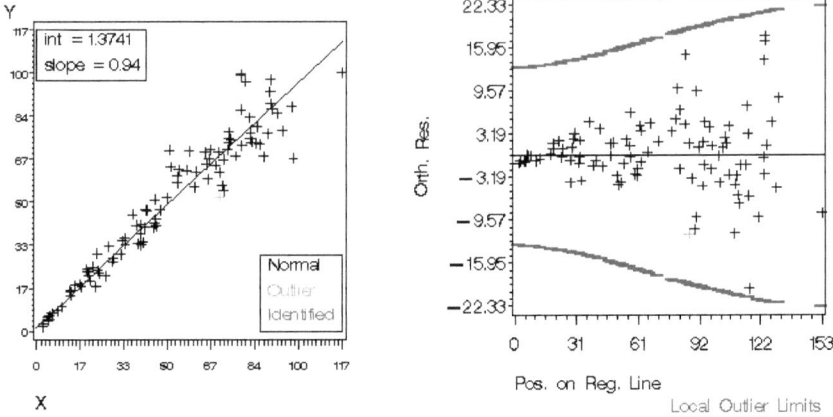

Figure B.36: Simulation 9: Constant Coefficient of Variance, One Outlier, Medium Outlier Term -The LORELIA Residual Test

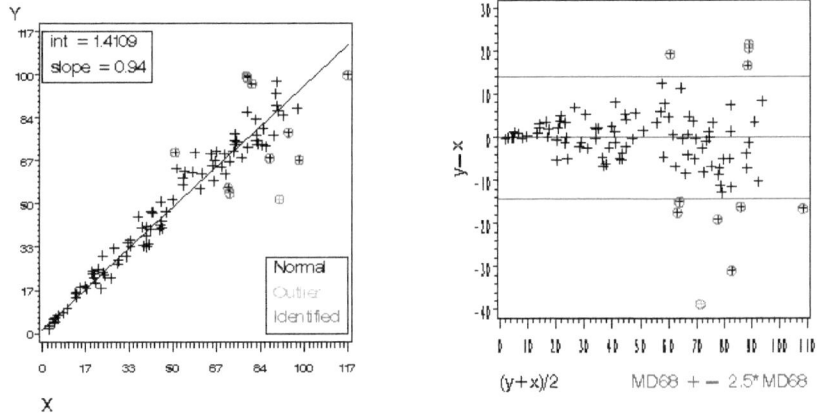

Figure B.37: Simulation 10: Constant Coefficient of Variance, One Outlier, High Outlier Term - Outlier Test for the Absolute Differences

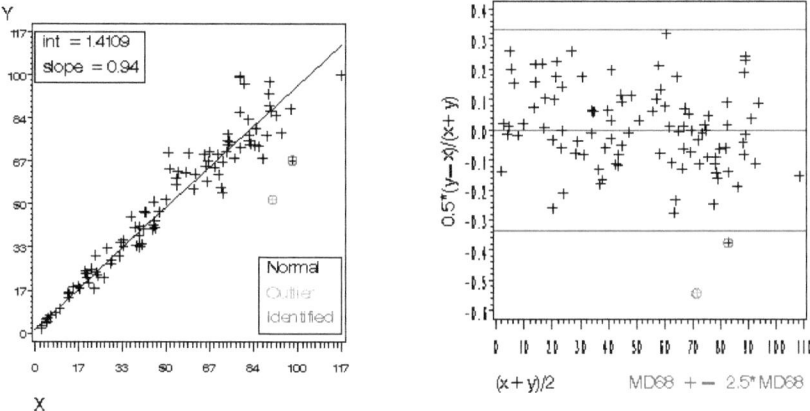

Figure B.38: Simulation 10: Constant Coefficient of Variance, One Outlier, High Outlier Term - Outlier Test for the Normalized Relative Differences

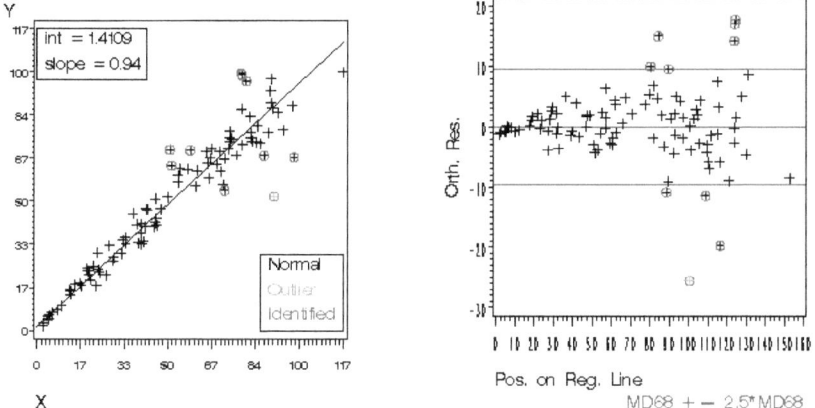

Figure B.39: Simulation 10: Constant Coefficient of Variance, One Outlier, High Outlier Term - Outlier Test for the Orthogonal Residuals

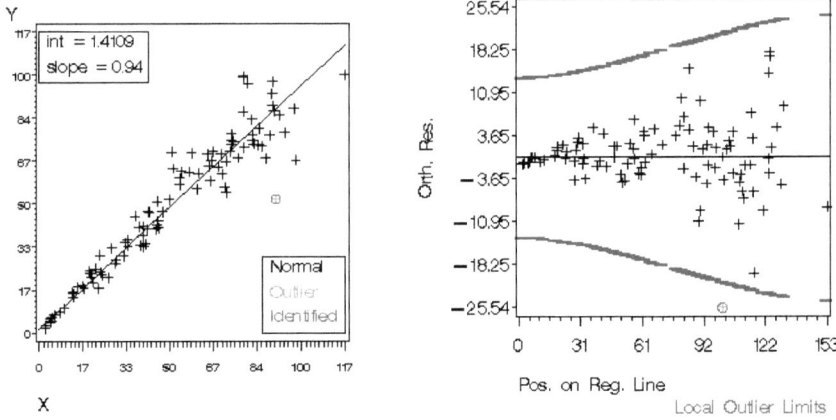

Figure B.40: Simulation 10: Constant Coefficient of Variance, One Outlier, High Outlier Term - The LORELIA Residual Test

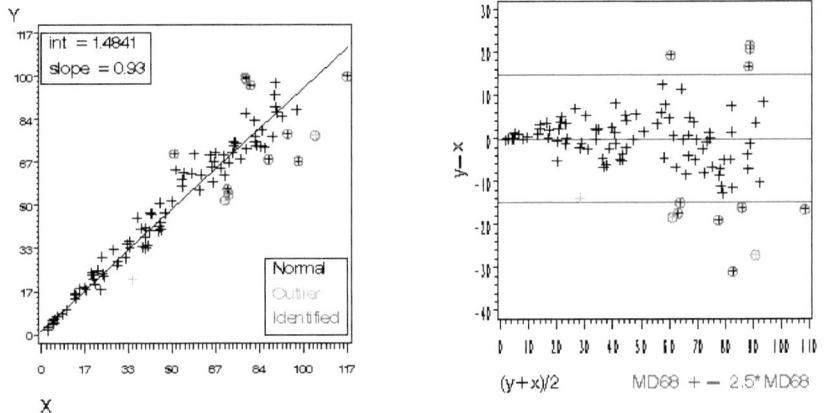

Figure B.41: Simulation 11: Constant Coefficient of Variance, Three Outliers, Medium Outlier Term, Uniformly Distributed Outliers - Outlier Test for the Absolute Differences

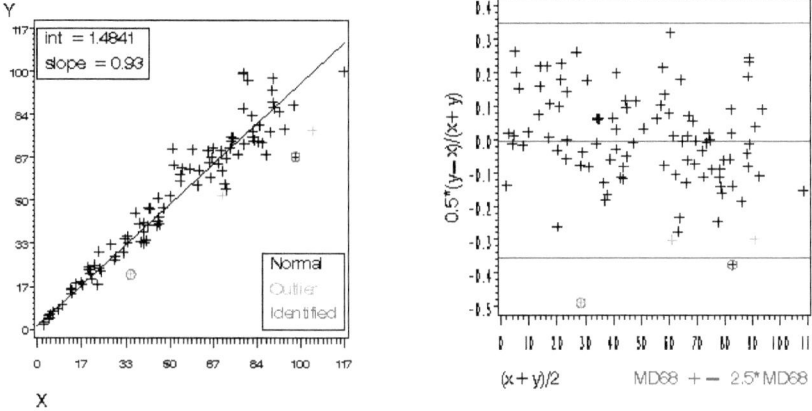

Figure B.42: Simulation 11: Constant Coefficient of Variance, Three Outliers, Medium Outlier Term, Uniformly Distributed Outliers - Outlier Test for the Normalized Relative Differences

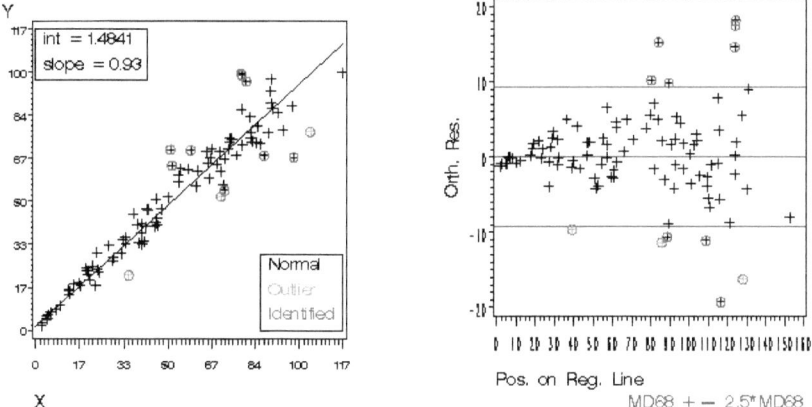

Figure B.43: Simulation 11: Constant Coefficient of Variance, Three Outliers, Medium Outlier Term, Uniformly Distributed Outliers - Outlier Test for the Orthogonal Residuals

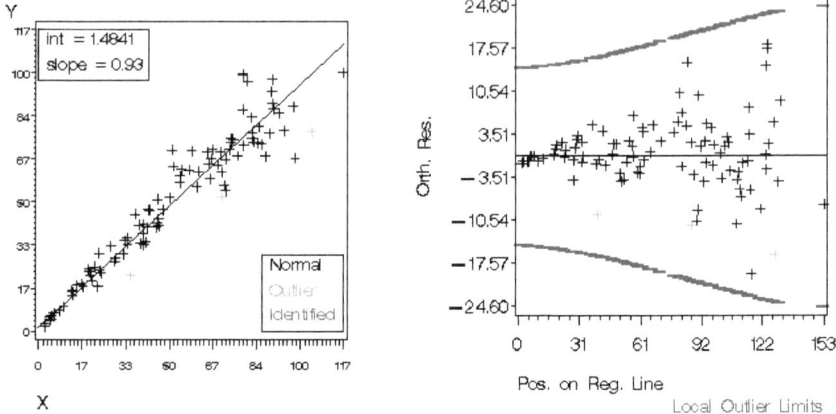

Figure B.44: Simulation 11: Constant Coefficient of Variance, Three Outliers, Medium Outlier Term, Uniformly Distributed Outliers -The LORELIA Residual Test

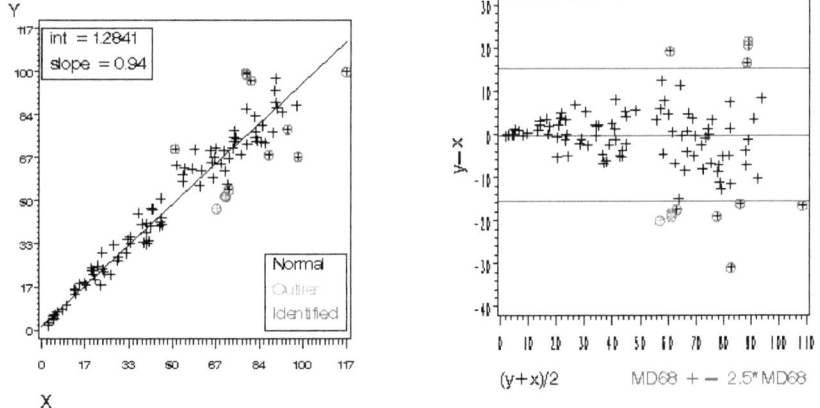

Figure B.45: Simulation 12: Constant Coefficient of Variance, Three Outliers, Medium Outlier Term, Clustered Outliers - Outlier Test for the Absolute Differences

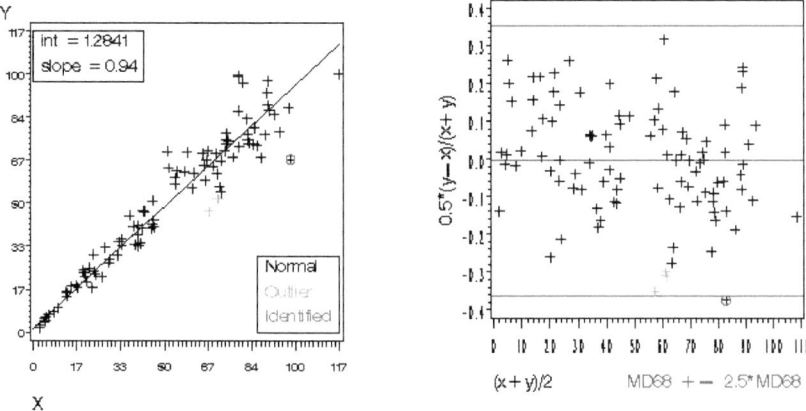

Figure B.46: Simulation 12: Constant Coefficient of Variance, Three Outliers, Medium Outlier Term, Clustered Outliers - Outlier Test for the Normalized Relative Differences

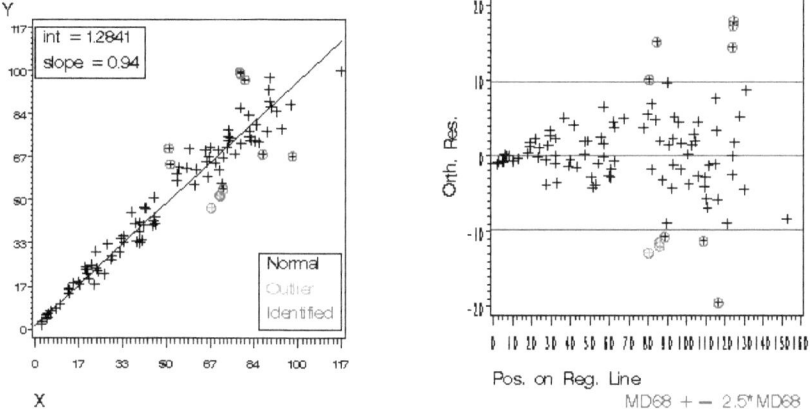

Figure B.47: Simulation 12: Constant Coefficient of Variance, Three Outliers, Medium Outlier Term, Clustered Outliers - Outlier Test for the Orthogonal Residuals

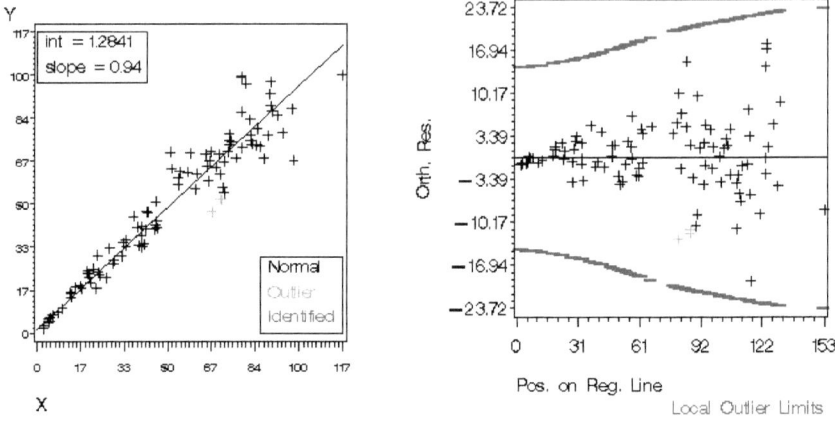

Figure B.48: Simulation 12: Constant Coefficient of Variance, Three Outliers, Medium Outlier Term, Clustered Outliers - The LORELIA Residual Test

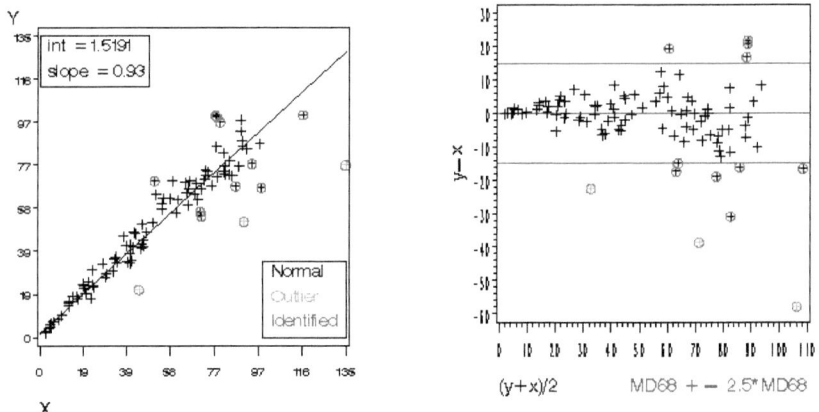

Figure B.49: Simulation 13: Constant Coefficient of Variance, Three Outliers, High Outlier Term, Uniformly Distributed Outliers - Outlier Test for the Absolute Differences

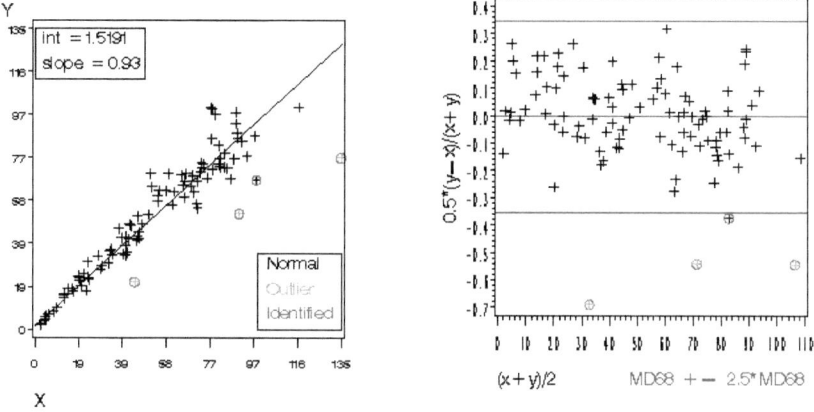

Figure B.50: Simulation 13: Constant Coefficient of Variance, Three Outliers, High Outlier Term, Uniformly Distributed Outliers - Outlier Test for the Normalized Relative Differences

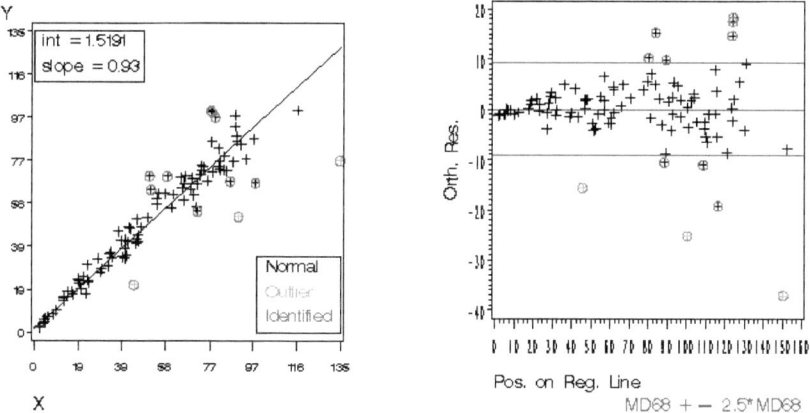

Figure B.51: Simulation 13: Constant Coefficient of Variance, Three Outliers, High Outlier Term, Uniformly Distributed Outliers - Outlier Test for the Orthogonal Residuals

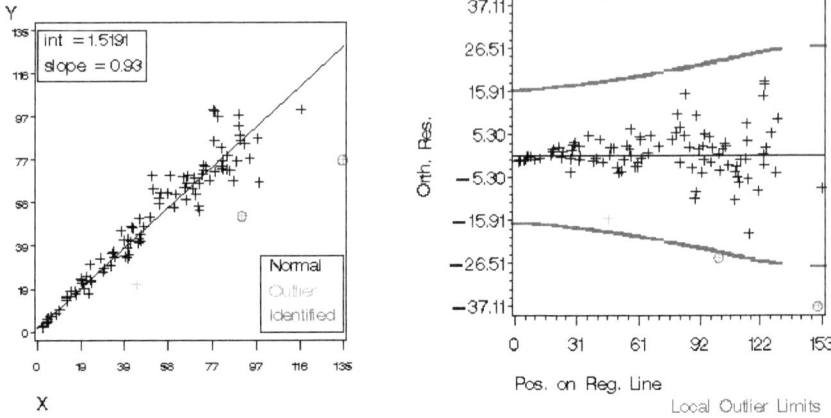

Figure B.52: Simulation 13: Constant Coefficient of Variance, Three Outliers, High Outlier Term, Uniformly Distributed Outliers -The LORELIA Residual Test

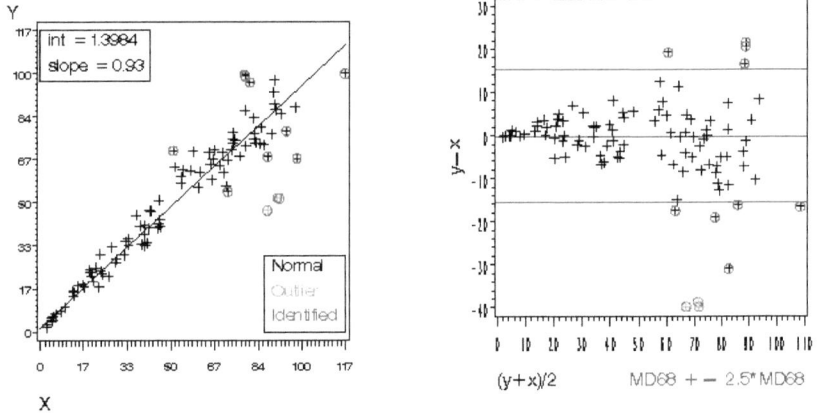

Figure B.53: Simulation 14: Constant Coefficient of Variance, Three Outliers, High Outlier Term, Clustered Outliers - Outlier Test for the Absolute Differences

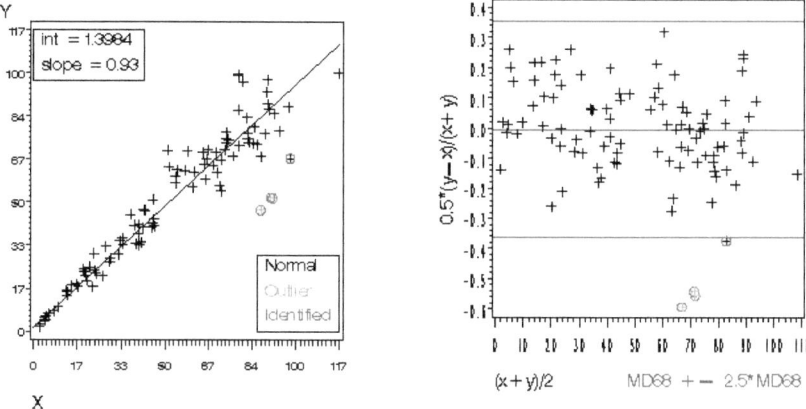

Figure B.54: Simulation 14: Constant Coefficient of Variance, Three Outliers, High Outlier Term, Clustered Outliers - Outlier Test for the Normalized Relative Differences

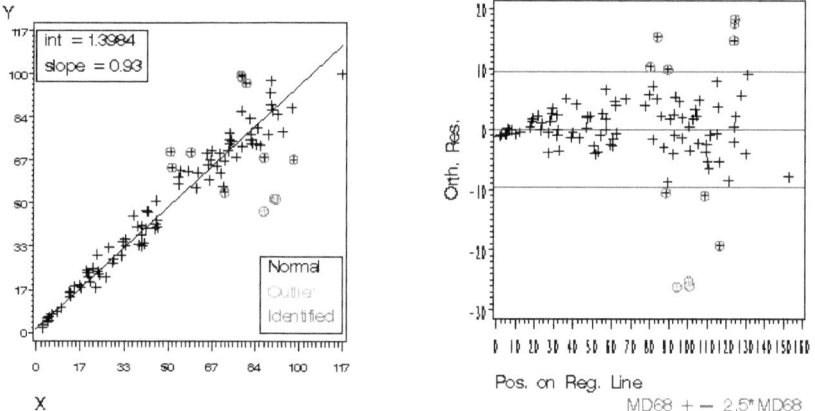

Figure B.55: Simulation 14: Constant Coefficient of Variance, Three Outliers, High Outlier Term, Clustered Outliers - Outlier Test for the Orthogonal Residuals

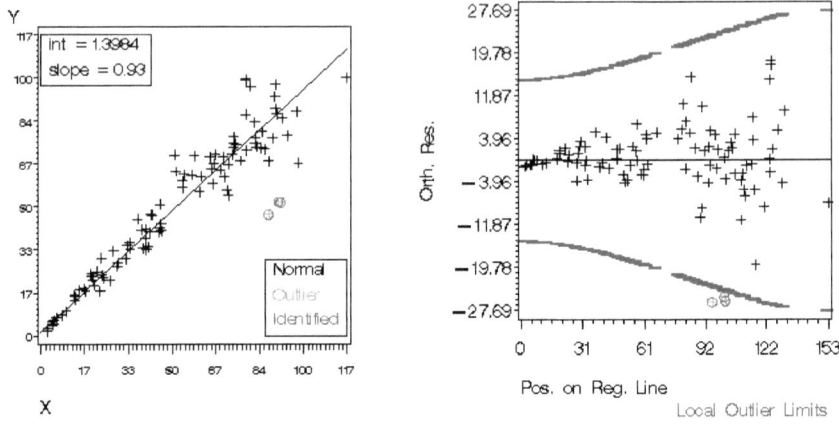

Figure B.56: Simulation 14: Constant Coefficient of Variance, Three Outliers, High Outlier Term, Clustered Outliers -The LORELIA Residual Test

B.3 Non Constant Coefficient of Variance

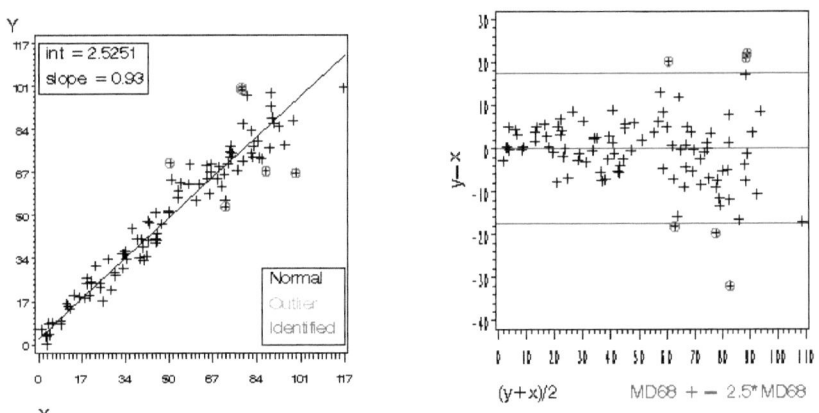

Figure B.57: Simulation 15: Non Constant Coefficient of Residual Variance, No Outliers - Outlier Test for the Absolute Differences

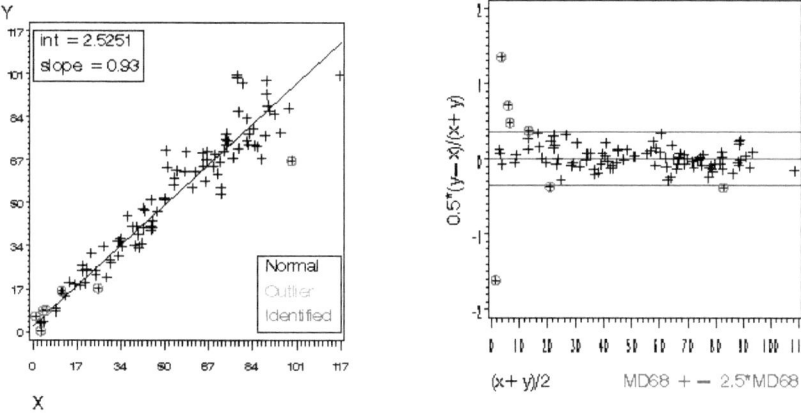

Figure B.58: Simulation 15: Non Constant Coefficient of Residual Variance, No Outliers - Outlier Test for the Normalized Relative Differences

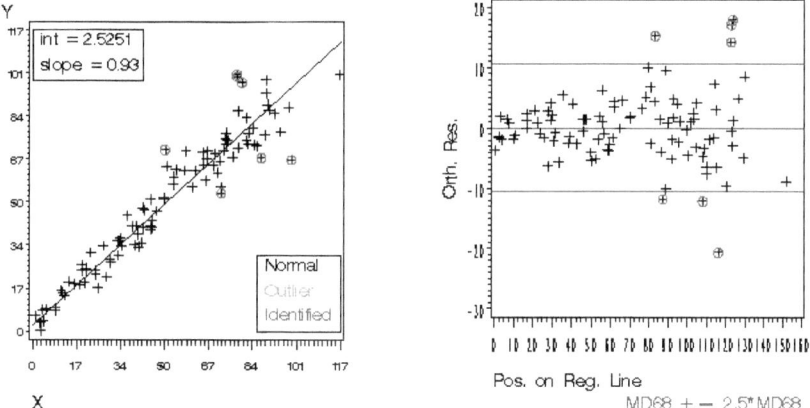

Figure B.59: Simulation 15: Non Constant Coefficient of Residual Variance, No Outliers - Outlier Test for the Orthogonal Residuals

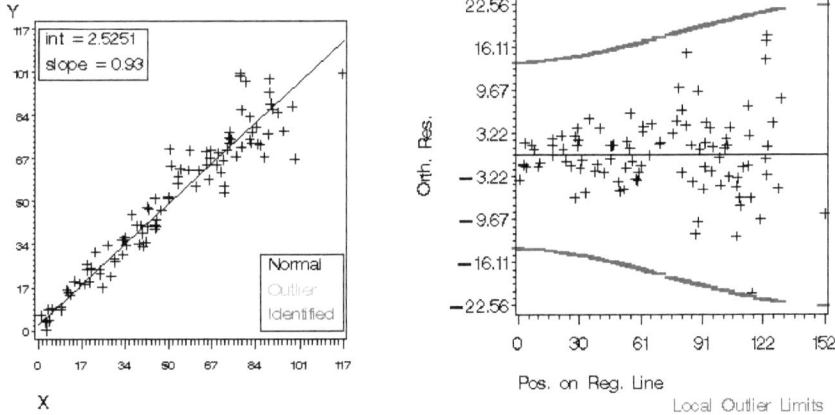

Figure B.60: Simulation 15: Non Constant Coefficient of Residual Variance, No Outliers - The LORELIA Residual Test

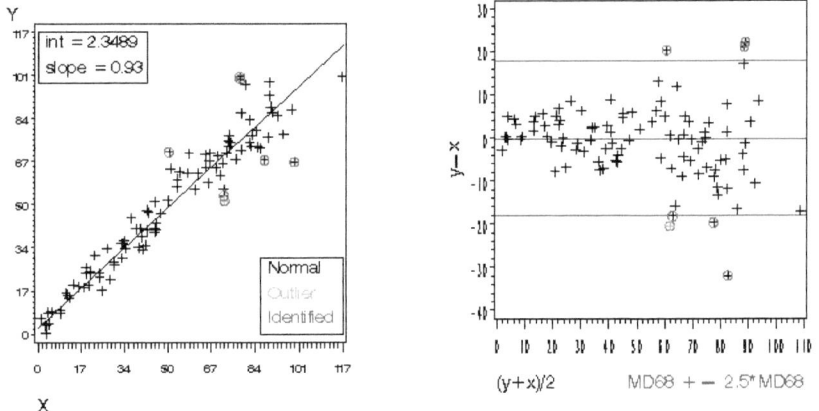

Figure B.61: Simulation 16: Non Constant Coefficient of Residual Variance, One Outlier, Medium Outlier Term - Outlier Test for the Absolute Differences

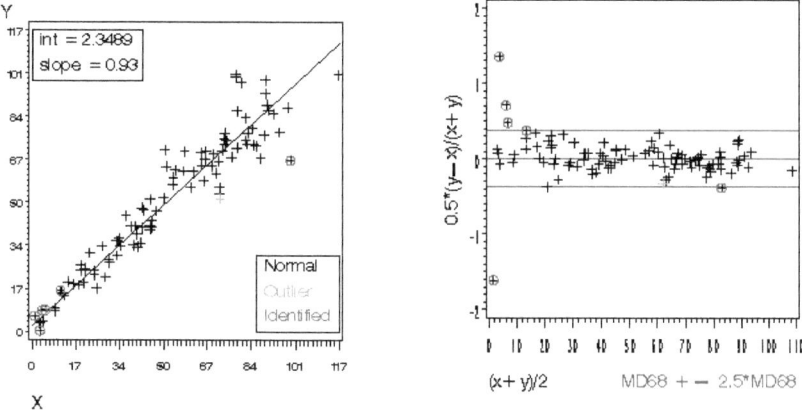

Figure B.62: Simulation 16: Non Constant Coefficient of Residual Variance, One Outlier, Medium Outlier Term - Outlier Test for the Normalized Relative Differences

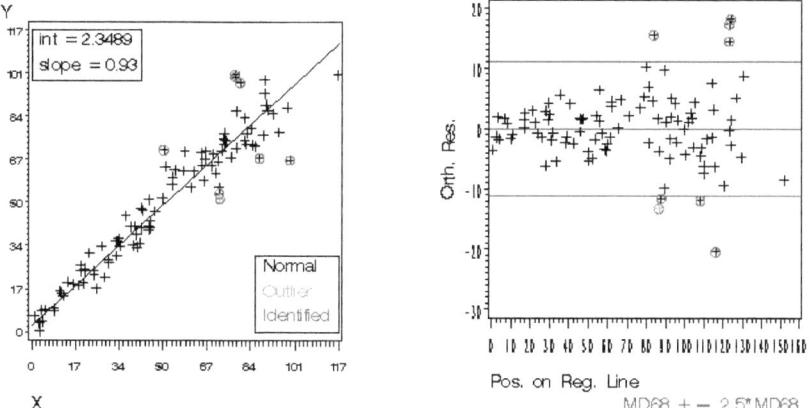

Figure B.63: Simulation 16: Non Constant Coefficient of Residual Variance, One Outlier, Medium Outlier Term - Outlier Test for the Orthogonal Residuals

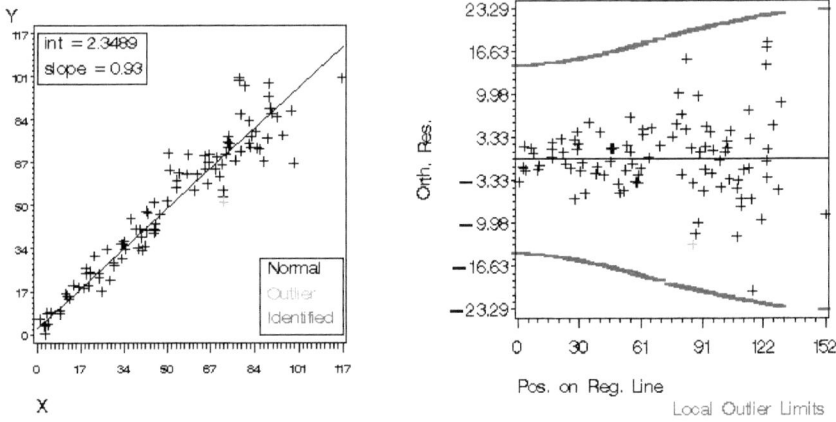

Figure B.64: Simulation 16: Non Constant Coefficient of Residual Variance, One Outlier, Medium Outlier Term -The LORELIA Residual Test

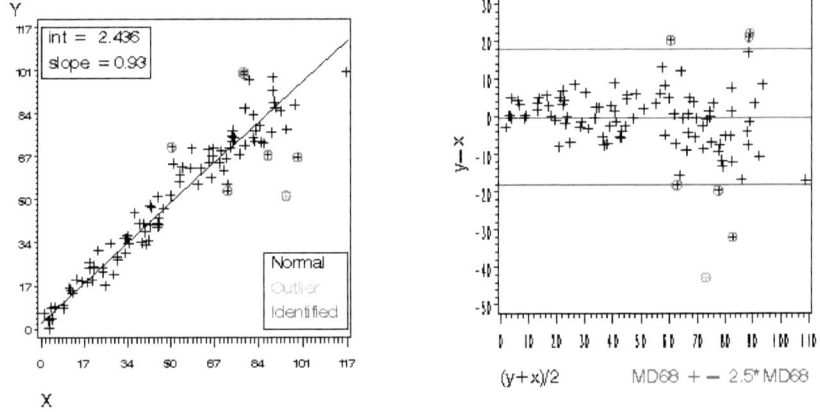

Figure B.65: Simulation 17: Non Constant Coefficient of Residual Variance, One Outlier, High Outlier Term - Outlier Test for the Absolute Differences

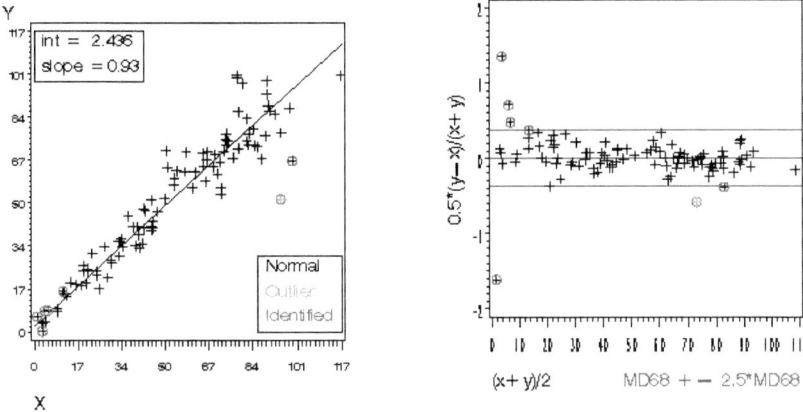

Figure B.66: Simulation 17: Non Constant Coefficient of Residual Variance, One Outlier, High Outlier Term - Outlier Test for the Normalized Relative Differences

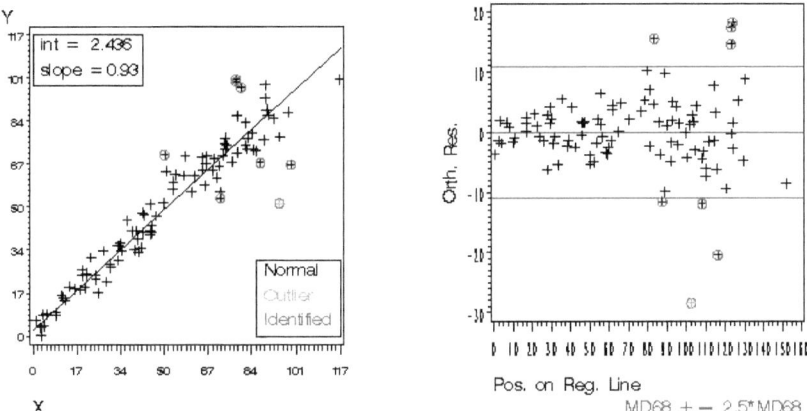

Figure B.67: Simulation 17: Non Constant Coefficient of Residual Variance, One Outlier, High Outlier Term - Outlier Test for the Orthogonal Residuals

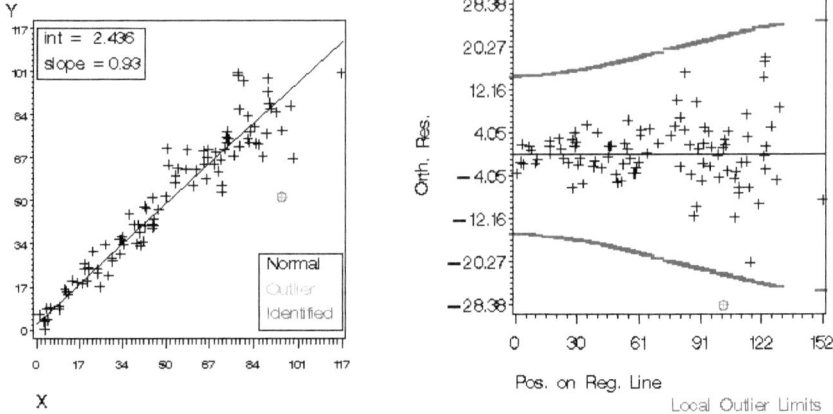

Figure B.68: Simulation 17: Non Constant Coefficient of Residual Variance, One Outlier, High Outlier Term -The LORELIA Residual Test

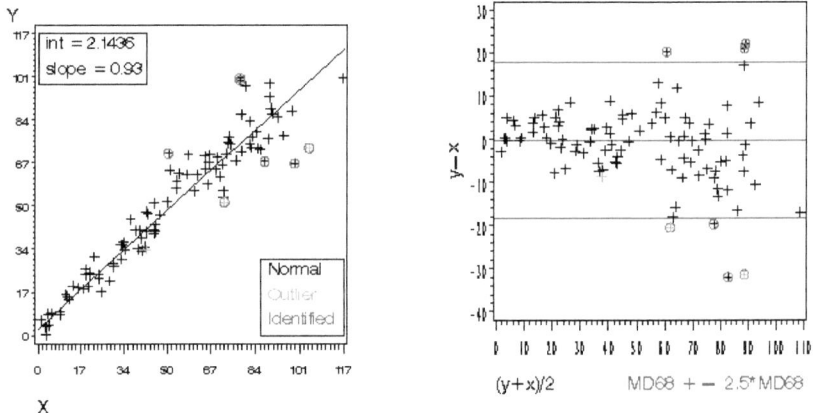

Figure B.69: Simulation 18: Non Constant Coefficient of Residual Variance, Three Outliers, Medium Outlier Term, Uniformly Distributed Outliers - Outlier Test for the Absolute Differences

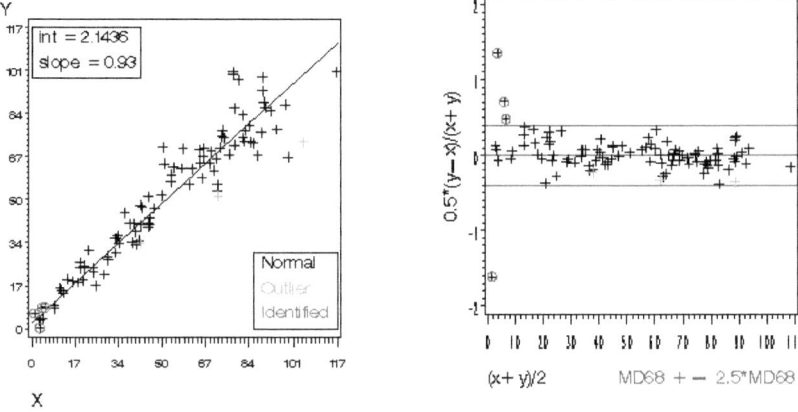

Figure B.70: Simulation 18: Non Constant Coefficient of Residual Variance, Three Outliers, Medium Outlier Term, Uniformly Distributed Outliers - Outlier Test for the Normalized Relative Differences

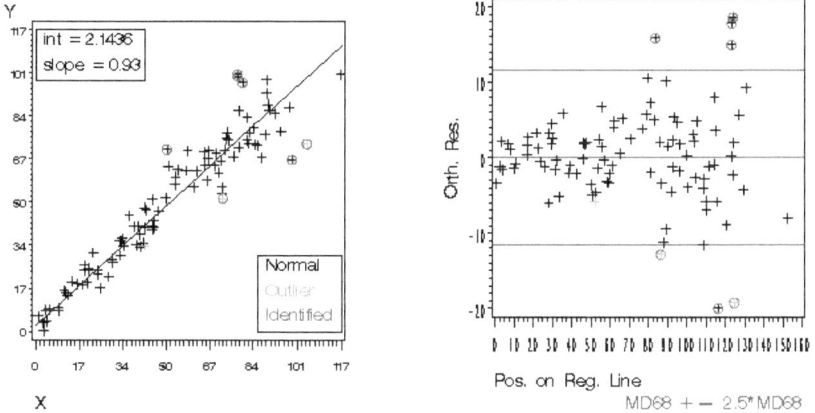

Figure B.71: Simulation 18: Non Constant Coefficient of Residual Variance, Three Outliers, Medium Outlier Term, Uniformly Distributed Outliers - Outlier Test for the Orthogonal Residuals

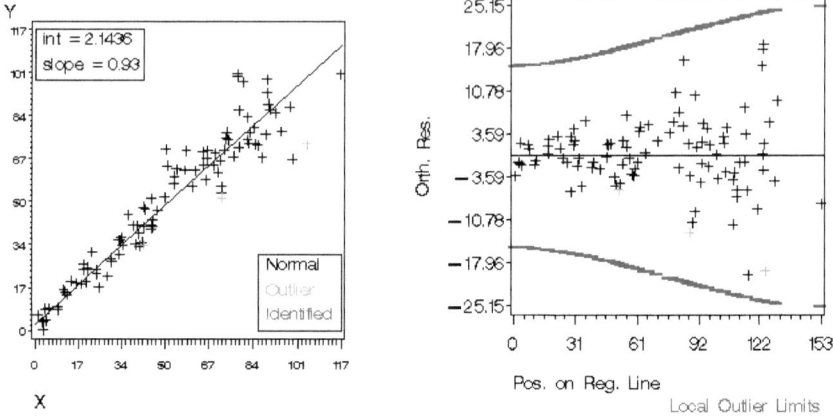

Figure B.72: Simulation 18: Non Constant Coefficient of Residual Variance, Three Outliers, Medium Outlier Term, Uniformly Distributed Outliers -The LORELIA Residual Test

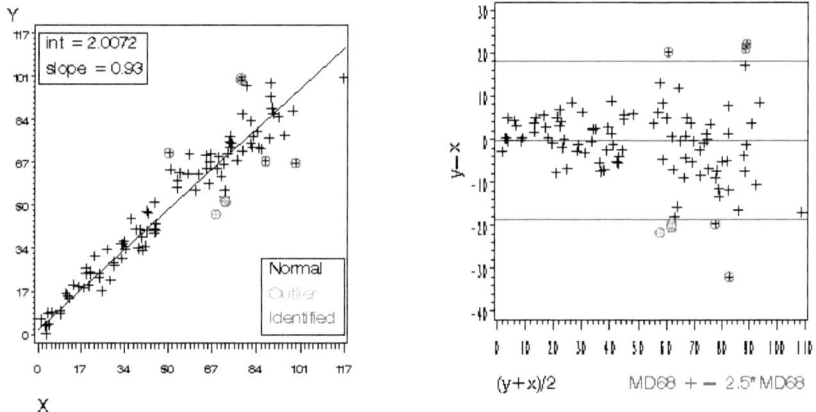

Figure B.73: Simulation 19: Non Constant Coefficient of Residual Variance, Three Outliers, Medium Outlier Term, Clustered Outliers - Outlier Test for the Absolute Differences

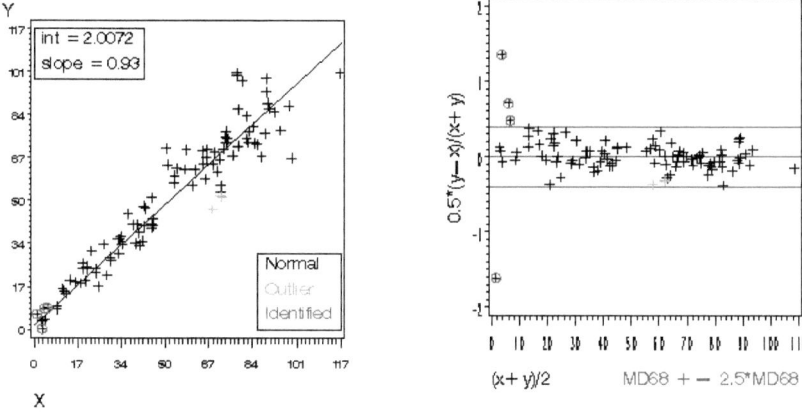

Figure B.74: Simulation 19: Non Constant Coefficient of Residual Variance, Three Outliers, Medium Outlier Term, Clustered Outliers - Outlier Test for the Normalized Relative Differences

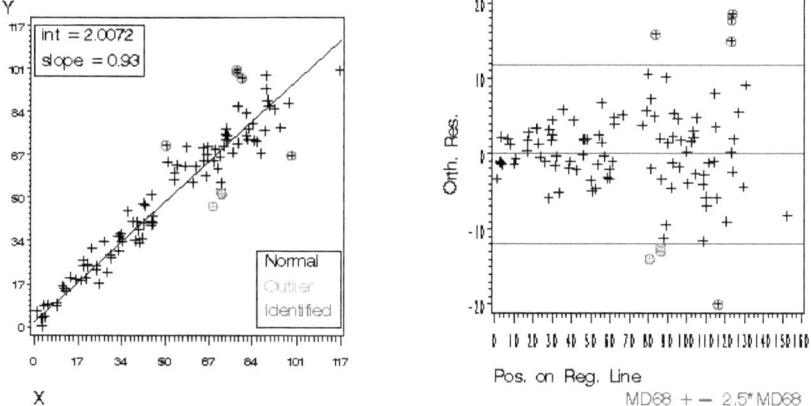

Figure B.75: Simulation 19: Non Constant Coefficient of Residual Variance, Three Outliers, Medium Outlier Term, Clustered Outliers - Outlier Test for the Orthogonal Residuals

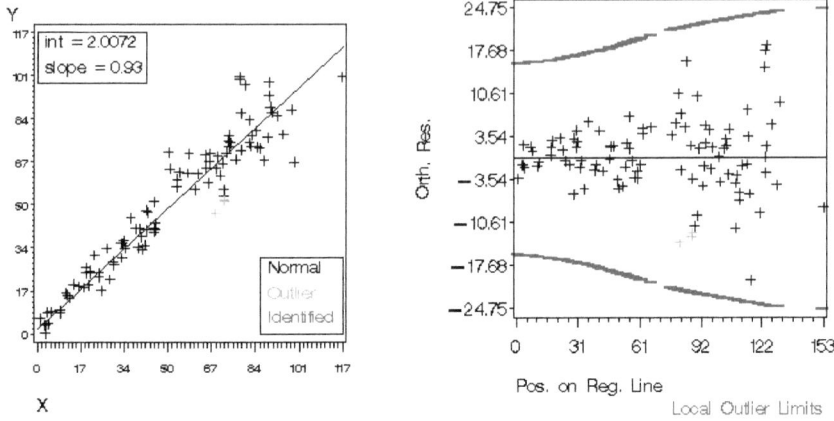

Figure B.76: Simulation 19: Non Constant Coefficient of Residual Variance, Three Outliers, Medium Outlier Term, Clustered Outliers -The LORELIA Residual Test

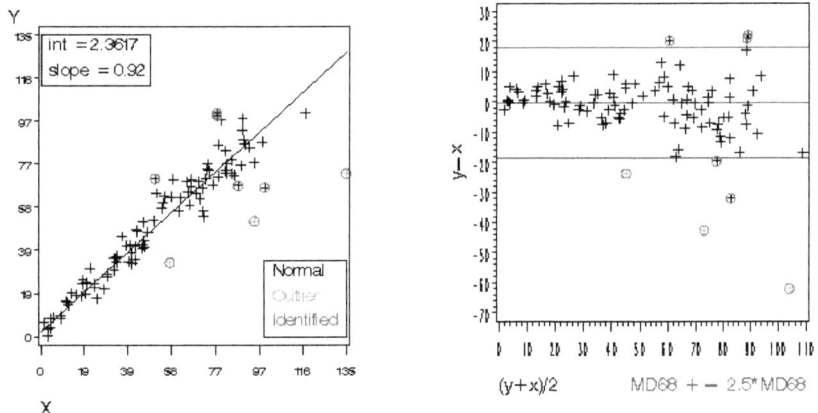

Figure B.77: Simulation 20: Non Constant Coefficient of Residual Variance, Three Outliers, High Outlier Term, Uniformly Distributed Outliers - Outlier Test for the Absolute Differences

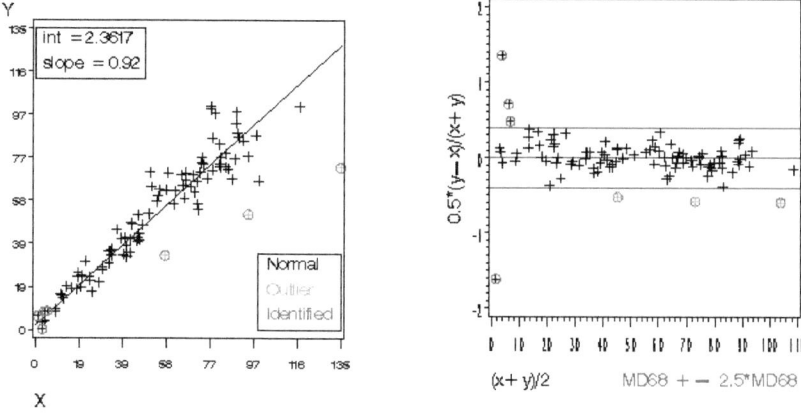

Figure B.78: Simulation 20: Non Constant Coefficient of Residual Variance, Three Outliers, High Outlier Term, Uniformly Distributed Outliers - Outlier Test for the Normalized Relative Differences

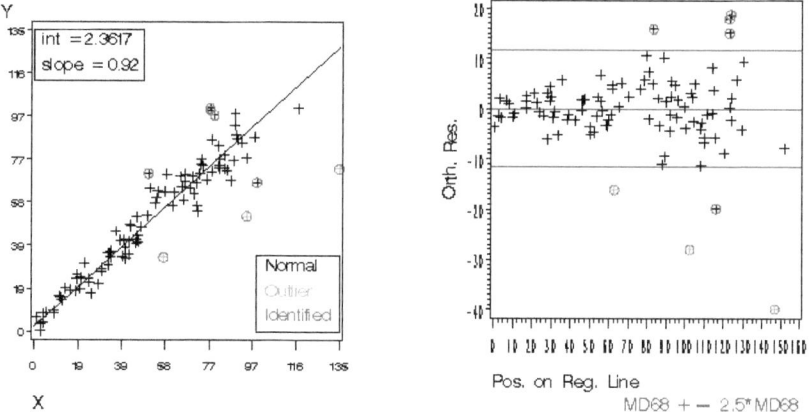

Figure B.79: Simulation 20: Non Constant Coefficient of Residual Variance, Three Outliers, High Outlier Term, Uniformly Distributed Outliers - Outlier Test for the Orthogonal Residuals

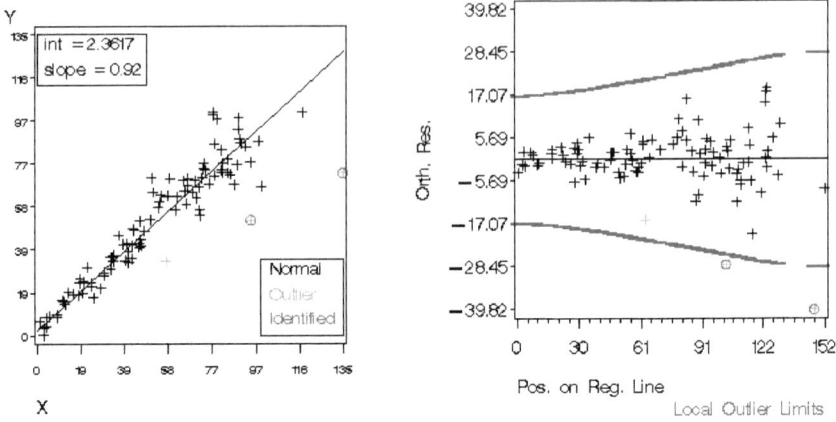

Figure B.80: Simulation 20: Non Constant Coefficient of Residual Variance, Three Outliers, High Outlier Term, Uniformly Distributed Outliers -The LORELIA Residual Test

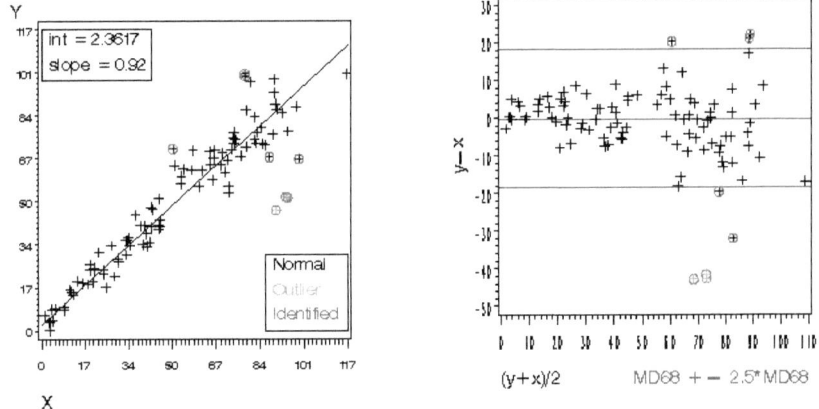

Figure B.81: Simulation 21: Non Constant Coefficient of Residual Variance, Three Outliers, High Outlier Term, Clustered Outliers - Outlier Test for the Absolute Differences

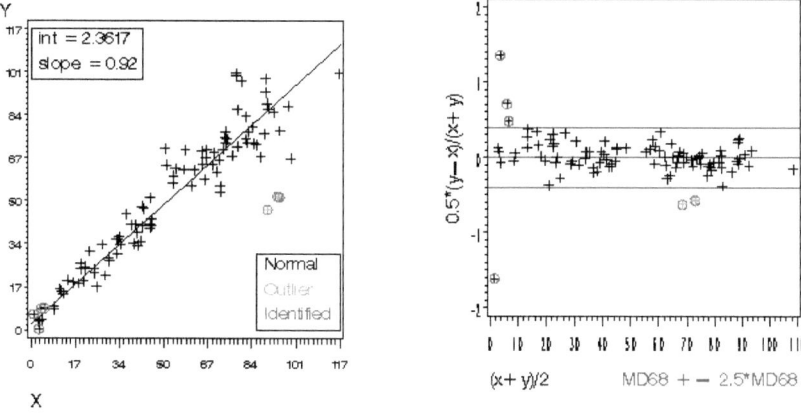

Figure B.82: Simulation 21: Non Constant Coefficient of Residual Variance, Three Outliers, High Outlier Term, Clustered Outliers - Outlier Test for the Normalized Relative Differences

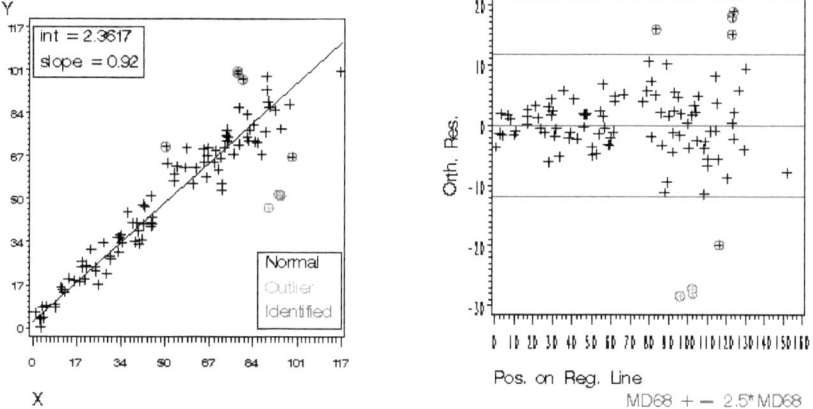

Figure B.83: Simulation 21: Non Constant Coefficient of Residual Variance, Three Outliers, High Outlier Term, Clustered Outliers - Outlier Test for the Orthogonal Residuals

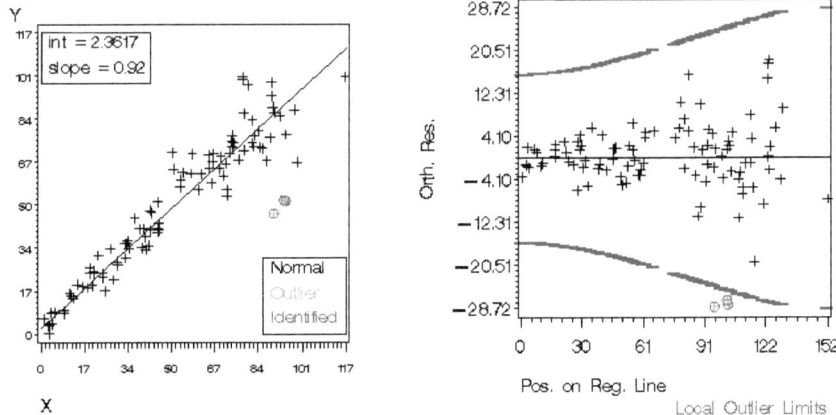

Figure B.84: Simulation 21: Non Constant Coefficient of Residual Variance, Three Outliers, High Outlier Term, Clustered Outliers -The LORELIA Residual Test

Symbols

$\mathbb{N}$	Natural Numbers
$\mathbb{R}$	Real Numbers
$\mathbb{R}_0^+$	Positive Real Numbers Including 0
$U(a,b)$	Continuous Uniform Distribution on $[a,b] \subset \mathbb{R}$
$N(\mu, \sigma^2)$	Normal Distribution with Expected Value μ and Variance σ^2
$logN(\mu, \sigma^2)$	Log Normal Distribution with Parameters μ and σ^2
χ^2_{DF}	χ^2 Distribution with DF Degrees of Freedom
$\Phi()$	Distribution Function of the Standard Normal Distribution
$z_{(1-\alpha)}$	$(1-\alpha)$ Quantile of the Standard Normal Distribution
$t_{DF,(1-\alpha)}$	$(1-\alpha)$ Quantile of the Students t Distribution with DF Degrees of Freedom
$X_1,...,X_n \stackrel{iid}{\sim} X$	$X_1,...,X_n$ are independent and identically distributed as X
$x_1,...,x_n$	Realizations of the Random Variables $X_1,...,X_n$
$x_{(1)},...,x_{(n)}$	Ordered Sequence of $x_1,...,x_n$
$E(X)$	Expected Value of the Random Variable X
$Var(X)$	Variance of the Random Variable X
$\overline{x}$	Mean Value of $x_1,...,x_n$ given by $\frac{1}{n}\sum_{i=1}^{n} x_i$
S^2_{xx}	Empirical Variance of X given by $\frac{1}{n-1}\sum_{i=1}^{n}(x_i - \overline{x})^2$
S^2_{xy}	Empirical Covariance of X and Y given by $\frac{1}{n-1}\sum_{i=1}^{n}(x_i - \overline{x})(y_i - \overline{y})$
med(x)	Median of $x_1,...,x_n$
mad68(x)	68% Median Absolute Deviation
$\min_{\{1 \leq i \leq n\}} \{x_i\}$	Minimum of $x_1,...,x_n$
$\max_{\{1 \leq i \leq n\}} \{x_i\}$	Maximum of $x_1,...,x_n$
$\inf_{\{1 \leq i \leq n\}} \{x_i\}$	Infimum of $x_1,...,x_n$
$\sup_{\{1 \leq i \leq n\}} \{x_i\}$	Supremum of $x_1,...,x_n$
sign()	Signum Function
mod()	Modulus Function
$\frac{\partial f}{\partial t}$	Partial Derivative of the Function f to t

H_0	Null Hypothesis of a Statistical Test
H_1	Alternative Hypothesis of a Statistical Test
α	Level of Significance
α_{loc}	Local Level of Significance for a Multiple Test Situation
α_{glob}	Global Level of Significance for a Multiple Test Situation
P_{int}	Population of Interest
P_{cont}	Contaminating Population
M_x, M_y	Methods which are to be compared
$\widetilde{X}, \widetilde{Y}$	Random Variables for the True Measurement Values of Methods M_x and M_y
E_x, E_y	Random Variables for the Measurement Errors in Methods M_x and M_y
$c_1, ..., c_n$	True Concentrations for Measurement Values $(x_1, y_1), ..., (x_n, y_n)$
$R_1, ..., R_n$	Random Variables for the Orthogonal Residuals
out_x, out_y	Outlier Term for Methods M_x and M_y
D_i^{abs}	Random Variable for the Absolute Difference between x and y
D_i^{rel}	Random Variable for the Relative Difference between x and y
D_i^{normrel}	Random Variable for the Normalized Relative Difference between x and y
$\widehat{\alpha}_{\text{PCA}}, \widehat{\beta}_{\text{PCA}}$	Parameter Estimators for Principal Component Analysis
$\widehat{\alpha}_{\text{SPCA}}, \widehat{\beta}_{\text{SPCA}}$	Parameter Estimators for Standardized Principal Component Analysis
$\widehat{\alpha}_{\text{PB}}, \widehat{\beta}_{\text{PB}}$	Parameter Estimators for Passing-Bablok Regression
R^2	Squared Correlation Coefficient for Linear Regression
C_α	$(1-\alpha)\%$ Approximative Confidence Intervall
(x_i^p, y_i^p)	Orthogonal Projection of the Measurement Tuple (x_i, y_i) to the Regression Line
w^{Shep}	Shepard's Weights (Inverse Distance Weights)
w^{Kon}	Weights proposed by [Konnert, 2005]
w_{ik}	LORELIA Weights
δ_{ik}	Squared Absolute Distance between (x_i^p, y_i^p) and (x_k^p, y_k^p)
Δ_{ik}	LORELIA Distance Weight, Transformation of δ_{ik}
$\gamma_{k,n}$	Reliability Measure
$\Gamma_{k,n}$	LORELIA Reliability Weight, Transformation of $\gamma_{k,n}$

Bibliography

[Acuña, Rodriguez, 2005] **Acuña, E., Rodriguez, C. (2005):** An Empirical Study of the Effect of Outliers on the Misclassification Error Rate. *Submitted to: Trans. Knowl. Data Eng.*

[Aitkin, Wilson, 1980] **Aitkin, M., Wilson, G. T. (1980):** Mixture Models, Outliers and the EM Algorithm. *Technometrics, Vol. 22, pp. 325 - 331.*

[Altman, Bland, 1983] **Altman, D. G., Bland, J. M. (1983):** Measurement in Medicine: The Analysis of Method Comparison Studies. *Statistician, Vol. 32, pp. 307 - 317.*

[Anscombe, 1960] **Anscombe, F. J. (1960):** Rejection of Outliers. *Technometrics, Vol. 2 , pp. 123 - 147.*

[Bablok et al. 1988] **Passing, H., Bablok, W., Bender, H., Schneider, B. (1988):** A General Procedure for Method Transformation - Application of Linear Regression Procedures for Method Comparison Studies in Clinical Chemistry, Part III. *J. Clin. Chem. Clin. Biochem., Vol. 26, pp. 783 - 790.*

[Barnett, Lewis, 1994] **Barnett, V., Lewis, T. (1994):** Outliers in Statistical Data. *Chichester: John Wiley & Sons.*

[Bernoulli, 1777] **Bernoulli, D. (1777):** Dijudicatio maxime probabilis plurium observationum discrepantium atque versimillima inductio inde formanda. *Acta Academiae Scientiorum Petropolitanae, Vol.1, pp. 3 - 33.*
English Translation by Allen, C.G. (1961), *Biometrica, Vol. 48, pp. 3 - 13.*

[Bland, Altman, 1986] **Bland J. M., Altman, D. G. (1986):** Statistical Methods for Assessing Agreement Between Two Methods of Clinical Measurement. *Lancet, Vol. 1, pp. 307 - 310.*

[Bland, Altman, 1995]	**Bland J. M., Altman, D. G. (1995)**: Comparing Methods of Measurement: Why Plotting Difference Against Standard Method is Misleading. *Lancet, Vol. 346, pp. 1085 - 1087.*
[Bland, Altman, 1999]	**Bland J. M., Altman, D. G. (1999)**: Measurement Agreement in Method Comparison Studies. *Stat. Meth. Med. Res., Vol. 8, pp. 135 - 160.*
[Box, Tiao, 1968]	**Box, G. E. P. (1968)**: A Bayesian Approach to some Outlier Problems. *Biometrica, Vol. 55, pp. 119 - 129.*
[Brown, 1988]	**Brown, M. (1982)**: Robust Line Estimation with Errors in Both Variables. *J. Amer. Statist. Assn., Vol. 77, pp. 71 - 79.*
[Burke, 1999]	**Burke, S. (1999)**: Missing Values, Outliers, Robust Statistics & Non-Parametric Methods. *LC•GC Europe Online Supplement, pp. 19 - 24.*
[Buttler, 1996]	**Buttler, M. (1996)**: Ein einfaches Verfahren zur Identifikation von Ausreißern bei multivariaten Daten. *Diskussionspapiere, Lehrstuhl für Statistik und Ökonometrie, Universität Erlangen, Vol. 9.*
[Chauvenet, 1863]	**Chauvenet, W. (1863)**: Method of Least Squares. *Appendix to Manual of Spherical and Practical Astronomie, Philadelphia: Lippincott, Vol. 2, Tables 593 - 599, pp. 469 - 566.*
[Cheng, Ness, 1999]	**Cheng, C.-L., van Ness, J. W. (1999)**: Statistical Regression with Measurement Error. *London: Arnold, Kendall's Library of Statistics 6.*
[Davies, Gather, 1993]	**Davies, L., Gather, U. (1993)**: Identification of Multiple Outliers. *J. Amer. Statist. Assn., Vol. 88, No. 423, pp. 782 - 792.*
[Deming, 1943]	**Deming, W. E. (1943)**: Statistical Adjustment of Data. *New York: John Wiley & Sons.*
[Dixon, 1950]	**Dixon, W. J. (1950)**: Analysis of Extreme Values. *Ann. Math. Statist., Vol. 22, pp. 68 - 78.*
[Doxygen, van Heesch, 2008]	**Van Heesch, D. (2008)**: Doxygen 1.5.8 User's Guide.
[Fahrmeir et al., 2007]	**Fahrmeir, L., Kneib, T., Lang, S. (2007)**: Regression - Modelle, Methoden und Anwendungen. *Berlin, Heidelberg: Springer Verlag.*

[Feldmann, 1992]   **Feldmann, U. (1992):** Robust Bivariate Errors-in-Variables Regression and Outlier Detection. *Eur. J. Clin. Chem. Clin. Biochem., Vol. 30, pp. 405 - 414.*

[Fishman, Moore, 1982]   **Fishman, G. S., Moore, L. R. (1982):** A Statistical Evaluation of Multiplicative Congruential Generators with Modulus $(2^{31} - 1)$. *J. Amer. Statist. Assn., Vol. 77, No. 1, pp. 29 - 136.*

[Fuller, 1987]   **Fuller, W. A. (1987):** Measurement Error Models. *New York: John Wiley & Sons.*

[Goodwin, 1913]   **Goodwin, H. M. (1913):** Elements of the Precision of Measurements and Graphical Methods. *New York: McGraw-Hill.*

[Grubbs, 1950]   **Grubbs, F. E. (1950):** Sample Criteria for Testing Outlying Observations. *Ann. Math. Statist., Vol. 21, pp. 27 - 58.*

[Guttman, 1973]   **Guttman, I. (1973):** Care and Handling of Univariate and Multivariate Outliers in Detecting Spuriosity - A Baysian Approach. *Technometrics, Vol. 15, pp. 723 - 738.*

[Haeckel,1993]   **Haeckel, R. (1993):** Evaluation Methods in Laboratory Medicine. *Weinheim: VCH Verlag.*

[Hartmann et al. 1996]   **Hartmann, C., Smeyers-Verbeke, J., Massart, D.L. (1996):** Detection of Bias in Method Comparison Studies by Regression Analysis. *Anal. Chim. Act., Vol. 338, pp. 19 - 40.*

[Hartmann et al. 1997]   **Hartmann, C., Vankeerberghen, P., Smeyers-Verbeke, J., Massart, D.L. (1997):** Robust Orthogonal Regression for the Outlier Detection when Comparing Two Series of Measurement Results. *Anal. Chim. Act., Vol. 344, pp. 17 - 28.*

[Hartung et al., 2009]   **Hartung, J., Elpelt, B., Klösener, K.-H. (2009):** Statistik - Lehr- und Handbuch der angewandten Statistik. *München: Oldenbourg.*

[Hawkins, 1980]   **Hawkins, D. M. (1980):** Introduction to Outliers. *London: Chapman & Hall.*

[Hawkins, 2002]   **Hawkins, D. M. (2002):** Diagnostics for Conformity of Paired Quantitative Measurements. *Statist. Med., Vol. 21, pp. 1913 - 1935.*

[Hochberg, Tamhane, 1987]   **Hochberg, Y., Tamhane, A. C. (1987):** Multiple Comparison Procedures. *New York: John Wiley & Sons.*

[Holm, 1979]   **Holm, S. (1979):** A Simple Sequentially Rejective Multiple Test Procedure. *Stand. J. Statist., Vol. 6, pp. 65 - 70.*

[Hsu, 1996]   **Hsu, J.C. (1996):** Multiple Comparisons - Theory and Methods. *New York: Chapmann & Hall / CRC.*

[Irwin, 1925]   **Irwin, J. O. (1925):** On a Criterion for the Rejection of Outlying Observations. *Biometrica, Vol. 17, pp. 238-250.*

[Konnert, 2005]   **Konnert, A. (2005):** Detection of Outliers in Method Comparison Studies. *Internal Report - Roche Diagnostics, Penzberg, Germany.*

[Linnet, 1998]   **Linnet, K. (1998):** Performance of Deming Regression Analysis in Case of Misspecified Analytical Error Ratio in Method Comparison Studies. *Clin. Chim., Vol. 44, No. 5, pp. 1024 - 1031.*

[Linnet, 1990]   **Linnet, K. (1990):** Estimation of the Linear Relationship between the measurements of two Methods with Proportional errors. *Statist. Med., Vol. 9, pp. 1463 - 1473.*

[Marks, Rao, 1979]   **Marks, R. G., Rao, P. V. (1979):** An Estimation Procedure for Data Containing Outliers with a One-Directional Shift in the Mean. *J. Amer. Statist. Assn., Vol. 74, pp. 614 - 620.*

[Olive, 2005]   **Olive, D. J. (2005):** Two Simple Resistant Regression Estimators. *Comp. Statist. Data Anal., Vol. 49, pp. 809 - 819.*

[Passing, Bablok, 1983]   **Passing, H., Bablok, W. (1983):** A New Biometrical Procedure for Testing the Equality of Measurements from Two Different Analytical Methods - Application of Linear Regression Procedures for Method Comparison Studies in Clinical Chemistry, Part I. *J. Clin. Chem. Clin. Biochem., Vol. 21, pp. 709 - 720.*

[Passing, Bablok, 1984]   **Passing, H., Bablok, W. (1984):** Comparison of Several Regression Procedures for Method Comparison Studies and Determination of Sample Size - Application of Linear Regression Procedures for Method Comparison Studies in Clinial Chemistry, Part II. *J. Clin. Chem. Clin. Biochem. Vol. 22, pp. 431-445.*

[Pearson, Sekar, 1936]   **Pearson, E. S., Sekar, C. C. (1936):** The Efficiency of Statistical Tools and a Criterion for the Rejection of Outlying Observations. *Biometrica, Vol. 28, pp. 308 - 320.*

[Peirce, 1852]   **Peirce, E. S. (1852):** Criterion for the Rejection of Doubtful Observations. *Astron. J., Vol. 2, pp. 161 - 163.*

[Qian, 1998]   **Qian, J. (1998):** Estimation of the Effective Degrees of Freedom in T-Type Tests for Complex Data. *Proceedings of the Survey Research Methods Section, ASA, pp. 704 - 708.*

[Rio et al., 2001]   **Del Rio, F. J., Riu, J., Rius, F. X. (2001):** Graphical Criterion for the Detection of Outliers in Linear Regression taking Account Errors in Both Axes. *Anal. Chim. Act., Vol. 446, pp. 489 - 494.*

[Rocke, Lorenzato, 1995]   **Rocke, D. M., Lorenzato, S.,(1995):** A two Component Model for Measurement Error in Analytical Chemistry. *Technometrics, Vol. 37, No. 2, pp. 176 - 184.*

[Rousseeuw, Leroy, 1987]   **Rousseeuw, P. J., Leroy, A. M.,(1987):** Robust Regression and Outlier Detection. *New York: John Wiley & Sons.*

[Rousseeuw, Zomeren, 1990]   **Rousseeuw, P. J., van Zomeren, B. C.,(1990):** Unmasking Multivariate Outliers and Leverage Points. *J. Amer. Statist. Assn., Vol. 85, No. 411, pp. 49 - 58.*

[SAS Insitute Inc., 2008]   **SAS Insitute Inc. (2008):** SAS® 9.1.3 User's Guide.

[Satterthwaite, 1941]   **Satterthwaite, F. (1941):** Synthesis of Variance. *Psychometrika, Vol. 6, No. 5, pp. 309 - 316.*

[Satterthwaite, 1946]   **Satterthwaite, F. (1946):** An Approximate Distribution of Estimates of Variance Components. *Biom. Bull., Vol. 2, pp. 110 - 114.*

[Shepard, 1968]   **Shepard, D. (1968):** A Two-Dimensional Interpolation Function for Irregularly-Spaced Data. *Proceedings of the 1968 ACM National Conference, pp. 517 - 524*

[Stökl et al., 1998]   **Stökl, D., Dewitte, K., Thienpont, L. M. (1998):** Validity of Linear Regression in Method Comparison Studies: Is it Limited by the Statistical Model or the Quality of the Analytical Input Data? *Clin. Chem., Vol. 44, No. 11, pp. 2340 - 2346.*

[Stone, 1868]   **Stone, E. J. (1868):** On the Rejection of Discordant Observations. *Monthly Notices Roy. Astr. Soc., Vol. 28, pp. 165 - 168.*

[Theil 1950]   **Theil, H. (1950):** *Proc. Kon. Ned. Akad. v. Wetensch. AS 3, Part I: pp. 386-392, Part II: pp. 521-525, Part III: pp. 1397 - 1412.*

[Thompson, 1935]	**Thompson, W.R. (1935):** On a Criterion for the Rejection of Observations and the Distribution of the Ratio of the Deviation to the Sample Standard Deviation. *Ann. Math. Statist., Vol. 6, pp. 214 - 219.*
[Tukey 1960]	**Tukey, J. W. (1960):** A Survey of Sampling from Contaminated Distributions. *In Olkin, I. (Editor) (1960): Contributions to Probability and Statistics, Standford, California: University Press.*
[Ukkelberg, Borgen, 1993]	**Ukkelberg, A., Borgen, O. S. (1993):** Outlier Detection by Robust Alternating Regression. *Anal. Chim. Act., Vol. 277, pp. 489 - 494.*
[Wellmann, Gather, 2003]	**Wellmann, J., Gather, U. (2003):** Identification of Outliers in a One-Way Random Effects Model. *Stat. Papers, Vol. 44, pp. 335 - 348.*
[Wadsworth, 1990]	**Wadsworth, H. M. Jr. (1990):** Handbook of Statistical Methods for Engineers and Scientists. *New York: McGraw-Hill*
[Wright, 1884]	**Wright, T.W. (1884):** A Treatise on the Adjustment of Observations by the Method of Least Squares. *New York: Van Nostrand*
[Xie, Wei, 2003]	**Xie, F.-C., Wei, B.-C. (2007):** Diagnostics Analysis for Log-Birnbaum-Saunders Regression Models. *Comp. Statist. Data Anal., Vol. 51, pp. 4692 - 4706.*

Die VDM Verlagsservicegesellschaft sucht für wissenschaftliche Verlage abgeschlossene und herausragende

Dissertationen, Habilitationen, Diplomarbeiten, Master Theses, Magisterarbeiten usw.

für die kostenlose Publikation als Fachbuch.

Sie verfügen über eine Arbeit, die hohen inhaltlichen und formalen Ansprüchen genügt, und haben Interesse an einer honorarvergüteten Publikation?

Dann senden Sie bitte erste Informationen über sich und Ihre Arbeit per Email an *info@vdm-vsg.de*.

Sie erhalten kurzfristig unser Feedback!

VDM Verlagsservicegesellschaft mbH
Dudweiler Landstr. 99
D - 66123 Saarbrücken
www.vdm-vsg.de

Telefon +49 681 3720 174
Fax   +49 681 3720 1749

Die VDM Verlagsservicegesellschaft mbH vertritt

Printed by Books on Demand GmbH, Norderstedt / Germany